AT THE DAWN OF THE SEXUAL REVOLUTION

AT THE DAWN OF THE SEXUAL REVOLUTION

Reflections on a Dialogue

IRA L. REISS AND ALBERT ELLIS

ALTAMIRA PRESS
A Division of Rowman & Littlefield Publishers, Inc.
Walnut Creek • Lanham • New York • Oxford

AltaMira Press
A Division of Rowman & Littlefield Publishers, Inc.
1630 North Main Street, #367
Walnut Creek, CA 94596
www.altamirapress.com

Rowman & Littlefield Publishers, Inc.
A Member of the Rowman & Littlefield Publishing Group
4720 Boston Way
Lanham, MD 20706

PO Box 317
Oxford
OX2 9RU, UK

Copyright © 2002 by AltaMira Press

All rights reserved. No part of this publication may be reproduced, stored in a retrieval system, or transmitted in any form or by any means, electronic, mechanical, photocopying, recording, or otherwise, without the prior permission of the publisher.

British Library Cataloguing in Publication Information Available

Library of Congress Cataloging-in-Publication Data
Reiss, Ira L.
   At the dawn of the sexual revolution : reflections on a dialogue / by Ira Reiss and Albert Ellis.
      p. cm.
   Includes bibliographical references.
   ISBN 0-7591-0272-4 (cloth : alk. paper)—ISBN 0-7591-0273-2 (pbk. : alk. paper)
   1. Sexual ethics. 2. Ellis, Albert—Views on sexual ethics. 3. Ellis, Albert—Correspondence. 4. Reiss, Ira L.—Views on sexual ethics. 5. Reiss, Ira L.—Correspondence. I. Ellis, Albert. II. Title.

HQ32 .R456 2002
176—dc21   2002001297

Printed in the United States of America

∞™ The paper used in this publication meets the minimum requirements of American National Standard for Information Sciences—Permanence of Paper for Printed Library Materials, ANSI/NISO Z39.48-1992.

Contents

Preface   vii
Ira L. Reiss

Preface   xi
Albert Ellis

CHAPTER 1

Abstinence Is Dethroned But What about Casual Sex?
Letters from July 7, 1956 to April 2, 1957   1

CHAPTER 2

Enter Ethics, Religion, and Publications: Letters from April 5, 1957 to September 7, 1957   51

CHAPTER 3

Therapeutic Ideas and the Launching of a Sexual Science Organization: Letters from November 10, 1957 to November 3, 1959   77

CHAPTER 4

New Books and New Institutes Blossom Forth: Letters from February 22, 1960 to October 4, 1967   123

CHAPTER 5

Overall Changes in My Views   163
Albert Ellis

CHAPTER 6

Taking Stock of My Past and Present Views   167
Ira L. Reiss

Appendix A: How I Became Interested in Sexology and Sex Therapy   175
Albert Ellis

Appendix B: My Path to Sexual Science   187
Ira L. Reiss

Preface

IRA L. REISS

MY LETTER TO ALBERT ELLIS in 1956 launched our debates about sexuality and other issues that were then surfacing in American society. The 1950s and 1960s were very important times in American sexual history. The cultural stage was being set for playing out the sexual revolution that fully burst forth at the end of this time period and continued well into the 1970s. Despite the pro-family and pro-religion aspects of the 1950s society, under the surface a sexual revolution was clearly being brewed. One of the key ingredients in producing this premarital sexual revolution was the loosening of parental control of children.

How did something like this happen in a pro-family and pro-religious era such as the 1950s? It happened because the women who had gone to work in American factories during World War II had permanently changed the traditional gender mold for women. They showed the country, and much more importantly, they showed themselves, that women could do any type of work and be just as good at it as men. Those experiences gave many "trapped housewives" a taste of freedom and accomplishment that whet their appetites for more of that elixir.

The most revealing statistic of social change was the sharp rise in the percentage of women employed who had preschool-age children from 1950 to today—increasing from a small minority to a majority. That was a crucial change because the nucleus of the traditional female gender role was the tie to small children. The loosening of this tie presaged a revolutionary change in both the female gender role and the upbringing of children. In 1960, in *Premarital Sexual Standards in America*, my first book, I predicted a sexual revolution that would start by the end of the 1960s. I then researched a national sample and several local samples and found strong support for my "youth and female autonomy explanation" of why our sexual standards were changing. I reported on this study in my second book (1967). In sum, I found that the 1950s and 1960s were a time when the seeds of greater female and youth autonomy were breaking through the resistant traditional soil of American society and producing a revolutionary change in our sexual standards.

As my ideas were surfacing, I looked for someone else to discuss them with. I chose Al Ellis because he was one of the most prolific and interesting authors who was critically analyzing American sexual customs. There was much we agreed on, but also much that we didn't, and the letters in this book make that all very clear. I feel that these letters are important in that they reveal the thinking at this crucial time of social change of two people who would write about sexuality for the next five decades and whose publications would be widely read by other professionals. Furthermore, the nature of society at that time is vividly revealed in these letters by our comments about sexuality, religion, gender, and family. Also, since Al is a psychologist and I am a sociologist, the letters focus on individual as well as social aspects of the world that was being primed to explode into an extraordinary sexual and gender revolution. Finally, for those readers with an "inquiring mind" the letters also reveal personal aspects of our lives.

Discovering these letters in my files was a very exciting event. I had been going over my extensive old files of letters in order to better understand my own career. I have learned from checking these original sources how much distortion time can bring to memory. So there was a real value in my having such letters—they forced me to examine today's memories in terms of other, less revisionist, documents from my past. Reading through these letters corrected some of my self-conceptions and gave me much more of the vivid colors and flavors of those years. These letters are presented to you the reader in their, so to speak, unadulterated form. Nothing has been changed or deleted and the characters are real.

I should mention that although all of Al's letters to me from the 1950s and 1960s are here, some of my letters to him are missing. I looked through all my files to try to locate them but evidently I was not as complete in saving my letters as I was in saving Al's letters to me. This was the precomputer, typewriter age, and making a copy of your letter entailed using carbon paper and watching out for smudges, and so I didn't do it every time. But most of my letters are here and the content of any missing letters is quite clear in the responses from Al. His letter style was to respond point by point to the issues I raised in my letters. So the exchange between us is always easily grasped. It is quite fitting that two people whose careers have both focused strongly on understanding sexuality and who have debated with each other their somewhat differing conceptions, should do this book together. The letters display the personal as well as the professional basis for our different but often overlapping ideas and feelings.

But we do much more than just present our letters in this book. Throughout the book, we present in detail how our early views have or have not changed over the past several decades. I begin the first four chapters expressing my ideas about our views then and now and Al puts forth his explanation of new perspectives in his commentaries presented right after many of his letters. In addition, each of us sums up our overall perspective on our thoughts and feelings "then and now" in our summary chapters that follow the four letter chapters. These two summary chapters attempt to understand the stability and the changes of the ideas and feelings expressed in our 1956–1967 letters over the decades since we wrote them. Do we still hold the same views? How and why have we changed? Just how predictive of future work and ideas were such early letters? And what is the relationship between the personal and the professional in our careers?

In order to better fill out the portrait, we have each included, at the back of the book, a one-chapter autobiography that we each recently wrote. These autobiographies will help the reader place our perspectives into our broader life/career context. This book will afford the reader a personal view that is very rarely seen in print of two well-known professional workers in the field of sexuality. Also, there is no shortage of sexual issues in today's society. I hope our thoughts here will be helpful to this generation of people struggling with our current sexual issues.

References

Reiss, Ira L. 1960. *Premarital Sexual Standards in America*. Glencoe, Ill.: Free Press.
———. 1967. *The Social Context of Premarital Sexual Permissiveness*. New York: Holt, Rinehart and Winston.

Preface

ALBERT ELLIS

I AM REALLY ENTHUSIASTIC about collaborating on this book with Ira Reiss. I had been doing some work on my autobiography before April 2000, in which I mentioned my early days in the field of human sexuality and my being influenced by the pioneering work of Ira and of several of my other friends, especially Henry Guze and Hans Lehfeldt. But I had partly forgotten about some of the details of Ira's correspondence with me on important sexual issues in the 1950s and 1960s. I was most pleasantly surprised when he sent me copies of our early correspondence that he had collected and put together and suggested that we publish them with our up-to-date commentaries. As I read through our letters, it took me remarkably few minutes to agree that they were well worth publishing. What a great idea!

The more I read of the Ellis–Reiss sheaf of letters, the more enthusiastic I became about seeing them in print. Not that I am a shy, retiring violet in the field of publishing. Hardly! I have authored almost seventy books and hundreds of articles—the proceeds of which go a long way to keep our busy psychotherapy institute alive and kicking. My reputation in the field of psychology is widespread. Do I really need to add an additional tome to my list? Not exactly.

As I read the Reiss-Ellis letters, however, several seemingly important reasons for getting them to press surfeited my mind. Here are a few of them:

1. The history of the dawn of the sexual revolution in the 1950s and 1960s has yet to be fully written. Our letters, scripted in fairly torrid heat at that time by two of its prime participants, provides a first-hand account of some of its fascinating aspects.
2. Several crucial issues in the field of human sexuality, most of which are still very much alive today, are discussed by us in a friendly but differing

manner, and proved to be influential on our own thinking and behavior. Also, I think I can say, they were potent on the ideas and practices of many people.
3. We showed in our letters both liberal and open-minded views about sex and love, as well as a liberal and open-minded acceptance of the theories and practices on which we personally disagreed. We accepted without liking many of each other's differences. We thereby showed our scientific and personal flexibility. I hope our book sets a model in this respect for others to follow.
4. We both learned considerably from our epistolary interchanges and grew and developed in the process. We distinctly held our ground but also mutually allowed the other to influence and change some of our philosophies of sex and love.
5. Because of our own openness to growth and development, I think we also significantly influenced many of our general readers to make sexual and amative advances.
6. I think that our hearing and accepting each other's views enabled us to change and develop in subsequent years and to radically change some of our own views on sex, sociality, psychotherapy, and other issues. In this book, we shall most specifically refer to some of these changes in some of our interpolated comments to our letters, as well as to summarize some of them in our postscripts.

For reasons such as these, I am enthusiastic about publishing these letters that Ira and I, with no premeditation or guile beforehand, now look forward to seeing in print. We shall see whether our readers agree!

Abstinence Is Dethroned But What about Casual Sex?
Letters from July 7, 1956 to April 2, 1957

Introduction by Ira L. Reiss

IN MY EARLY ARTICLES, I was writing about our society's bias against premarital sex. Al's early writings also expressed a critique of American sexual values. I sensed our compatible value orientation and sent him a reprint of my first journal article on the double standard in premarital sexuality (1956, 224–230). He wrote back thanking me and that started our long correspondence. Back in 1956, most writers were stressing multiple reasons for saving sex for marriage and it was a relief to find someone who shared my more pluralistic approach to sexuality. Al was even more acceptant of premarital sexuality than I was in that he did not emphasize affectionate sex more than casual sex. Although I accepted casual sex, I saw affectionate sexuality as much superior and I wrote Al asking him why he seemed to make less of a distinction than I did between what I called "body-centered sexuality" and "person-centered sexuality" (1957, 334–338). Thus began a series of letters on this relative ranking and related issues. That debate is central in many of the letters in this first chapter. Al eventually referred to our correspondence on this topic in his 1958 book *Sex without Guilt*. My own writing plans at this time involved finishing writing my first book *Premarital Sexual Standards in America*, in which I was dealing extensively with the differences between casual (body-centered) and affectionate (person-centered) sexuality and how each type fit into our society. Both of our professional writing plans supported the extensive letter exchange on this topic.

As the reader will see in these early letters, one of the reasons for Al's stronger endorsement of body-centered sexuality was that his explanatory schema gave greater weight than mine to a biological sex drive. As he said in his January 25, 1957 letter: "I firmly believe that sex is a biological, as well as a social drive, and that in its biological phases it is essentially non-affectional." He did believe that person-centered sex was better but since people like body-centered sex too, he felt we should not blame ourselves when we pursue that. I indicated in my February 2, 1957 letter that I felt that body-centered sex is acceptable but that someone who concentrates on this type of sex is like a person who reads "comic books instead of more worthwhile literature." I posited a kind of Gresham's Law where bad sex would drive out good sex. And so I favored only occasional casual sex that did not disrupt the focus on affectionate sexual relationships. I do believe that my stress on the social and not the biological nature of sexuality was stronger than his and this was a key reason for our difference.

Al's sexual value scale seemed to utilize a quantitative hedonistic calculator and I was adding a qualitative element of affection that was hard to quantify. At this stage in our thinking, Al was the freer, more acceptant person. He had a wider ranging pluralism and more of a minimalist approach to ethics. There also was a difference on this issue based on Al being a psychologist and I being a sociologist. He was more sensitive to individual differences and I was more sensitive to social and cultural influences and pressures. Although we both had a strong acceptance of individual sexual choice, I was more interested in ranking these choices in terms of how they fit with our basic social values of affection, equality, honesty, and responsibility. Later, we will each comment on whether these differences have held up over the years.

I do mention in my March 18, 1957, letter that "[p]erhaps we've been fighting straw men—you fighting a position of 'blame and guilt' and I fighting a position of 'no preferred type of coitus'. Whereas in reality neither of us held these positions." But the letters did reveal some real differences. I sum up those differences in my February 13, 1957 letter: "I value coitus with affection more than you do and value coitus without affection less than you do." Nevertheless, we were both very strongly critical of the conflict ridden way that American society dealt with sexuality and of the double standard favoring male sexuality. On that we fully agreed. We ended this series of letters by arranging to meet when my wife Harriet and I were scheduled to be in New York at the Eastern Sociological Society meetings in April 1957. After writing all those letters, we were both rather curious about each other and looking forward to our first meeting.

ALBERT ELLIS, PH. D.
PARC VENDOME
333 WEST 56TH STREET
NEW YORK 19, N. Y.

July 7, 1956

Dr. Ira L. Reiss
Dept. of Sociology
College of William & Mary
Williamsburg, Virginia

Dear Dr. Reiss:

Thank you for sending me a reprint of your paper on the double standard in premarital sexual intercourse.

I feel that you have done an excellent, objective job of handling a ticklish subject, and shall look forward to reading your future papers in this field.

With best wishes,

Sincerely yours,

Albert Ellis

ALBERT ELLIS, PH. D.
PARC VENDOME
333 WEST 56TH STREET
NEW YORK 19, N. Y.

January 25, 1957

Dr. Ira L. Reiss
Dept. of Sociology & Anthropology
College of William & Mary
Williamsburg, Va.

Dear Dr. Reiss:

I can understand your being puzzled by some of my writings on sex and affection, largely because I have never quite considered the problem, in print, in long enough detail. My views are these:

I would agree with you that it is often DESIRABLE that an association between coitus and affection exist—particularly in marriage, since I feel that it is often quite difficult for two individuals to keep finely tuned to each other sexually over a period of years, and if there is not a good deal of love between them, they will tend to feel sexually imposed upon by the other. But the fact that the coexistence of sex and love may be DESIRABLE does not, to my mind, make it NECESSARY. My reasons for this view are several:

1. Many individuals DO find great satisfaction in sex relations without love; and I do not consider it fair to label these individuals as criminal just because they may be in the minority. I doubt very much, as you imply, that such persons are rare; I am sure, in fact, that they number literally millions.

2. Even individuals who find greater satisfaction in sex-love relations than in sex-sans-love relations rarely do so for ALL their lives. During much of their existence, especially the younger years, they often tend to find sex-without-love quite satisfying. When they become older, and the sex drives tend to wane, they may well emphasize sex with rather than without affection. But why should they be condemned WHILE they still prefer sex to sex-love affairs?

3. Many of the individuals, especially females in our culture, who say that they only enjoy sex when it is accompanied by affection are actually being unthinkingly conformist, and often unconsciously hypocritical. If they were courageous and unbiased, they would often enjoy sex without love. And why should they not?

4. I firmly believe that sex is a biological, as well as a social, drive, and that in its biological phases it is essentially non-affectional. If this is so, then we can expect that, however we try to civilize the sex drives—and civilize them to SOME degree we certainly must—there will always tend to be an underlying tendency for them to escape from our society-inculcated shackles and to be felt in the raw. When so felt, I do not see why we should make their experiencers feel guilty.

5. Many individuals—many millions, I am afraid—have little or no capacity for affection for love. The majority of these individuals, perhaps, are emotionally disturbed, and should preferably be helped to gain in their affectional propensities. But a large number are not particularly disturbed, but are just neurologically or cerebrally deficient. Morons and borderline subnormals, for example, are notoriously shallow in their feelings, and probably intrinsically so. Since these individuals—both the neurotic and the organically deficient-are for the most part, in our day and age, NOT going to be properly treated and NOT going to overcome their deficiencies, and since most of them very definitely DO have sex desires, I again see no point in making them guilty when they have non-loving sex relations.

6. Under some circumstances—these, I admit, probably ARE rare—some people find considerably more satisfaction in non-loving sex relations even though, under other circumstances, these SAME people may find more satisfaction in sex-love relations. Thus, the man who NORMALLY enjoys copulating with his wife because he loves as well as is sexually attracted to her, may occasionally greatly enjoy copulating with another woman with whom he has a purely animal-like, non-loving, and at times even semi-sadistic affair. Granting that this may be unusual, I do not see why it should be condemnable.

7. If many people can get along excellently and most cooperatively with business partners, employees, professors, laboratory associates, acquaintances, and others for whom they have little or no basic love or affection, but with whom they have certain specific things in common, I fail to see why there cannot be a good many people who can get along excellently and most cooperatively with sex mates with whom they may have little else in common. I personally would think it quite tragic if I ever spent much time with a girl with whom I had nothing in common but sex. I would also think it still relatively tragic if I spent much time with a girl with whom I had sex, friendship, cultural interests, etc., but no love interest. This is because I would like, in my 70-odd years of life, to have maximum satisfactions with any female with whom I spend much time. But I can—and have—spent a LITTLE time with many girls with whom I had common sex and cultural but little love interests; and I feel that, for the time I expended, my life was immeasurably enriched thereby. Moroever, if I were to become enamored for a time, and spend considerable time and effort thinking about and being with, a girl who largely obsessed me sexually, and for whom I had little or no love, I would merely view this as one of the penalties of my being a fairly normal biosocial human being who is easily disposed to that kind of behavior. I think, in other words, that the

great tragedy of human sexuality is that it endows us, and perhaps particularly us males, with a propensity to become quite involved and infatuated with females with whom, had we no sex urges, we would have little or nothing else in common. That is too bad; and it would probably be a better world if it were otherwise. But it is <u>not</u> otherwise, and I think it silly and pernicious for us to condemn ourselves because it is so. We had better, I think, ACCEPT our biosocial tendencies, or our fallible humanity, to some degree—instead of constantly blaming ourselves for it, and futilely trying to change certain of its aspects.

For reasons such as these, therefore, I feel that although it is usually—though not always—DESIRABLE for human beings to have sex relations with those they love rather than with those they do not love, it is by no means NECESSARY that they do so. And when we teach that it IS necessary, we only needlessly condemn literally millions to self-blame and atonement.

I hope that this explains my views, and would be most interested in hearing your reactions to them.

Sincerely,

Albert Ellis

Comments by Albert Ellis on His January 25, 1957 Letter

I think this is a good answer to Ira, and I endorse nearly all of it today. It advocates democracy, pluralism, and nonabsolutism in sex, love, and other behavior—which I and Rational Emotive Behavior Therapy (REBT) importantly uphold today.

I say in this letter that sex, in its biological phase, is essentially nonaffectional. I would correct this today, since I have stated in my first paper on REBT at the annual convention of the American Psychological Association in Chicago in August 1956 that thinking, feeling, and behaving are not disparate but are all intrinsically and "holistically integrated." Therefore, I was careless in this letter when I stated that the "biological"—or presumably behavioral—phase of sex is "nonaffectional." Biologically, sex is behavioral *and* affectional—that is, physical *and* emotional. It includes *desire* which is cognitive-emotional. Also, as John Bowlby has shown, affection and love are biologically *as well as* socially learned. In fact, social learning itself has strong biological foundations.

I refer in this letter to women as "girls." This was conventional usage in 1957, but it is sexist and I have not used the term "girls" for about twenty years, except to designate very young females. Today, I usually refer to a man's "girlfriend" as his "womanfriend."

February 2, 1957

Dear Dr. Ellis:

I have just finished spending a week grading over 100 final examination papers of my students. I'm convinced that I am much more exhausted than they are. However, I do think that answering your letter will be a refreshing way to spend part of the afternoon.

First let me say that although I do feel that there are significant differences between our views, I also think that in many ways we are quite similar. I have no doubt that there are millions of people who enjoy coitus without affection in fact, I would venture to say that most people would. When I stated that I could accept this coitus without affection only in rare cases—I did not mean that it was pleasureful only in rare cases or to rare individuals, I meant that I thought it could be <u>justified</u> only in relatively rare cases even though it would bring pleasure, in many more instances. I would also agree that sex has a biological side to it and that this side has no controls regarding affectionate requirements. I would also agree that because of this any sexual standard will incur violations. Again the fact that there are neurotic and biologically deficient individuals who could not achieve coitus with affection also seems reasonable to me, although I was not thinking of such atypical groups. With all of these comments of yours I would agree.

I think perhaps where we would disagree is in our attitudes towards coitus without affection and its consequences. I personally feel that such intercourse is sorely lacking—lacking in the sense that one can gain so much more from coitus with affection that even from a purely hedonistic viewpoint it seems advisable to seek that sort of coitus. The fact that people can gain satisfaction from coitus without affection does not seem to me to be justification for it. I would imagine that many professional criminals gain satisfaction from their work and perhaps dictators enjoy their behavior also; but surely the fact that someone enjoys an action is not in itself justification for that action. One must look at the total consequences of the behavior and in the case of coitus without affection I think the total context proves to be wanting.

Let me elaborate my criticism—I do not look on coitus without affection as something "sinfull", or "wrong" in the traditional meaning of these terms. I look at someone who concentrates on this type of outlet as I do upon someone who reads comic books instead of more worthwhile literature or someone who could eat nourishing food but stuffs himself with candy instead. In short, I feel that the person is immature and is missing out on greater enjoyments because of his or her desire to satisfy their immediate impulses.

I would agree that one cannot always be extremely fond of another person and thus there will be many times when if one indulges only in coitus with affection he or she will have to abstain. However, it seems to me that we have to abstain from pleasures in all walks of life and to try and avoid all pain is to try and avoid life. Of course, one should not use this analysis to justify pain, especially needless pain. However, in this case it seems to me that if one engaged in coitus without affection when they were not emotionally involved there would be a very real danger of this type of coitus becoming habitual and displacing coitus with affection. One might note that coitus without affection involves less personal restraints and is easier to obtain and thus it could tend to become habitual. Also PW/OA conflicts with our notion of love and encourages adultery. Sort of like "Greshams law," "bad sex driving out good sex." Also, even if this did not occur it is possible that one would form conflicting associations with sexual intercourse—in one case it is an act between two people who are strongly affectionately involved with each other and are expressing this fact in their sexual behavior—in another case it is an expression of bare sexual desire. The two connotations seem to conflict. Carrying out this line of reasoning I would say that if one could occasionally engage in coitus without affection and it would not weaken or disrupt his adherence to coitus with affection, but would merely act as a "change of pace", then I would have no objection to such behavior.

I do not think we are too far apart, for you said in your last letter that you tend to avoid concentrating on coitus without affection and do concentrate on coitus with affection. Our main difference seems to be in why we hold to this position and in our feelings about people who do otherwise, I think also we differ on the area of the place of biological drives in man's existence. You seem to feel that man should not inhibit these drives and feel almost that it is impossible to control them anyhow. I think that man's social nature is of crucial import and that what we call sexual drives exists only in theory—in practice man has a drive that is so mixed with his cultural training and experiences that it is just as social as is the "drive" that may impel him to control it. I think therefore that man can be trained to act sexually anyway at all and although I would agree that the closer we adhere to <u>some</u> form of available satisfaction, rather than to abstinence, the easier the standard will be to abide by, I still see no necessity to have a standard that guarantees immediate sources of satisfaction. There are many types of pleasure and I think one of our human searches is to find those that are more worthwhile and to be able to follow them, once we find them.

Before I wander around anymore I best summarize and close—I feel that coitus with affection leads to worthwhile consequences which are largely lacking in coitus without affection. I feel that in order to obtain these consequences it is *worthwhile to give up the pleasures and satisfactions that may be obtained from coitus without affection, I, of course, allow for exceptions as stated above, but no more than this, even though such restrictions may lead to guilt feelings and/or inhibitions of a sort.*

Now that we've both sort of stated our feelings it ought to be easier to see just where our difference is. This topic has always interested me and is part of a manuscript I am now working on—so I would be most interested to hear from you about your views.

Sincerely,

Dept of Soc.
College of Wm & Mary
Williamsburg, Va.

P.S. Another characteristic of coitus without affection just occurred to me. Such intercourse involves only a minute part of one's personality as compared to coitus with affection and this is perhaps the basis for it being less able to satisfy one in many ways. However, my thought here is that because it involves only part of one's personality and is given so much importance, it is quite possible that it may lead to behavior that is unacceptable to the rest of one's personality and thus create much conflict rather than relieve conflict as one might think such a pleasurable activity would. For example, one might break an appointment with a close friend because he has just met a sensuous female and even though the other girl means more to him emotionally he may follow his sexual desires. Again, some people lie about their religion, occupation, and so forth in order to impress someone who they only care for in a physical fashion. Granting that the more common cultural background one has and the more affection involved the less this occurs— it seems that short of a relatively strong dose of affection these drawbacks are quite possible, although of course, in varying degrees.

ALBERT ELLIS, PH. D.
PARC VENDOME
333 WEST 56TH STREET
NEW YORK 19, N. Y.

Feb. 8, 1957

Dr. Ira L. Reiss
Dept. of Sociology
College of Wm. & Mary
Williamsburg, Va.

Dear Dr. Reiss:

Let me take up the points in your last letter seriatim, so that we can see where we agree and disagree.

1. When you say that coitus without affection is biologically pleasureable in many instances but can be <u>justified</u> only in a few, you could mean that it is (a) wicked and/or (b) self-defeating. You claim that you really only mean (b) and not (a), and I will accept that. I would then object, however, that "justified" is a poor choice of word, since it inevitably, in our culture, has connotations of "just" and "right" as opposed to "unjust" and "wrong".

2. Let us, for the moment, forget about point (a) and stick to (b). You say that a person who has coitus without affection "is immature and is missing out on greater enjoyments because of his or her desire to satisfy their immediate impulses." But this would only be true if such an individual (whom we shall assume, for the sake of discussion, WOULD get greater enjoyment from love with sex) were USUALLY or ALWAYS having non-affectionate coitus. If he were OCCASIONALLY or SOMETIMES having love with sex, and the rest of the time having sex without love, he would be missing out on very little, if any, pleasure.

Under these circumstances, in fact, he would normally get MORE pleasure from SOMETIMES having sex without love. For the fact remains, and must not be unrealistically ignored, that in our culture, at the present time, sex without love is MUCH MORE FREQUENTLY available than sex with love. Consequently, to ignore non-affectional coitus when affectional coitus is NOT AVAILABLE would, it seems to me, be sheer folly. In relation both to immediate AND greater enjoyment, the individual would be losing out.

The claim can of course be made that if an individual gives up sex without love NOW he will experience much more pleasure by having sex with love IN THE FUTURE. This is an interesting claim; but I see no empirical evidence to sustain it. In fact, on theoretical grounds it seems most unlikely that it ever will be sustained. It is akin to the claim that if an individual starves himself for several days in a row he will greatly enjoy his meal at the end of a week or a month. I am sure he will—providing that he is then not too sick or debilitated to enjoy anything. But, even assuming that he derives enormous satisfaction from his one meal a week or a month, is his TOTAL satisfaction thereby greater than it would have been if he had enjoyed three good meals a day for the whole period of time? I distinctly doubt it.

So with sex. Anyone who has starved himself sexually for a period of time—as virtually every individual who rather rigidly sticks to the sex with love doctrine must—will (perhaps) ULTIMATELY achieve greater satisfaction, when he does find sex with love, than he would have had he been sexually freer. But, even assuming that this is so, will his TOTAL satisfaction be greater—particularly in view of the fact that he may have sex with love for perhaps one-sixth of his adult life and sex without love for five-sixths?

3. There is no evidence to warrant the belief, which you posit, that non-affectional sex would drive out affectional sex, somewhat in accordance with Gresham's law of currency. On the contrary, there is much reason to believe that just because an individual kept having sex, for quite a period, on a non-affectional basis, he would be more than eager to replace it with sex with love. From my clinical experience, I have often found that the males in our culture who most want to settle down, in a socalled "mature" manner, to having a single mistress or a wife, are those who have tried numerous lighter sex affairs and found them wanting. Your view that sex without love would become "habitual" and therefore lead to a lack of need for sex with love is an interesting hypothesis—but I have no reason for supposing that the facts, if obtainable, would support it. It is somewhat akin to the "ignorance is bliss theory"—which I have never found works very well in practice. For you are as much as saying that if people never experienced sex with love they would never realize how good it was and therefore would not strive for it. Or else you are saying that sex without love is so greatly satisfying, and sex with love such an intrinsic difficult and disadvantageous thing, that given the choice between the two, most people would pick the former. If this is so, then by all means let them pick the former; I think they would be, in terms of their GREATER and TOTAL happiness, be much better off. I doubt, however, whether your hypothesis IS factually sustainable and can only repeat that I have clinically found that individuals who are capable of sex with love usually wind up with it; while those who stick with non-affectional sex either are not particularly capable of sex with love or really WOULD be not especially better off with it.

As a psychotherapist, I will grant you, of course, that we often take individuals who because of their emotional difficulties are not capable of certain behavior, such as having affectional sex relations, and "cure" them so that they later are so capable.

But this does not give us reason to infer, as I think you seem to be inferring, that ALL persons who are incapable of, or who take no great pleasure in, sex without love are "neurotic and biologically deficient." I have reason to believe, from my clinical works and my biographical readings, that many quite worthwhile human beings, such as Immanuel Kant for example, were so dedicated to some form of activity that they were not interested in or did not have time for involved affectional relations. I see no reason why such individuals should not have perfectly normal sex urges, and why they should not consummate their urges on a non-affectional basis in order to have more time and energy for their other chosen pursuits.

I not only contend that many individuals in our culture, particularly males, can easily "occasionally engage in coitus without affection and . . . not weaken or disrupt his adherence to coitus with affection," but I would put this hypothesis much more strongly and hold that many such individuals can *often* engage in non-affectional coitus without weakening or disrupting their adherence to affectional coitus. I have known many such persons, both as friends and patients; and I am sure that many more, literally millions more, presently exist.

4. I agree with you that man's sex drives are biosocial or biocultural rather than merely biological; and I do not feel, as you state in your last letter "that man should not inhibit these drives and . . . that it is impossible to control them anyhow." I believe that man definitely SHOULD inhibit his sex drives in many instances; that this is necessary to cooperative, social living. I believe, for example, that it is entirely unethical and immoral for a man to pretend that he loves a woman, and hence get her into bed. I think, that, at the expense of his sex drives, he should be ruthlessly honest with himself and his sex partners, so that minimal hurt or ego-destruction may thus ensue. But I believe that any sane society, if one ever existed (which, to my knowledge, has never been the case), would put MINIMAL and NECESSARY restraints, rather than MAXIMUM and UNNECESSARY ones, on the sex drives—just as I believe that any sane society should put MINIMAL and NECESSARY restraints on man's urge to talk, write, or vote freely. People can certainly NEEDLESSLY hurt others by slander, wire-tapping, voting several times in one election, etc.; and I think they should be restrained from doing so. They can also NEEDLESSLY hurt others sexually and should be restrained or, preferably, learn to restrain themselves when there is a danger of their doing so. But I fail to see why their HARMLESS biosocial drives, such as (in my opinion) their drive to have non-affectional sex relations with willing partners when affectional ones are not available or would be too disadvantageous to arrange, should ARBITRARILY be bottled up because some post-Christian apologist THINKS they should be.

I heartily agree with you that, within some minor limits, "man can be trained to act sexually any way at all"; but I fail to see why he SHOULD be when the training essentially consists of blackmailing him into inhibiting quite harmless, exceptionally pleasurable drives. If it could be unequivocally demonstrated that forcing virtually all men and women into a sex-should-accompany-affection mold WOULD make for greater human happiness and mental health, I would be all for this kind of training. My thesis, however, is just the opposite: that anything but a sex-MAY-for-SOME-people-be PREFERABLE-with-rather-than-without-affection rule will almost certainly lead to LESS happiness and MORE emotional disturbance. I should greatly like to see an empirical test of our differing hypotheses, but I am afraid that any critical experiment on the subject that might be suggested would hardly be supported by the Ford Foundation.

5. I don't doubt that you are correct in assuming that coitus without affection involves relatively little of one's personality—or one's attitudes towards oneself and others—than does coitus with affection. For this very reason, however, I would say that in a sane society it would involve much LESS self-conflict than does coitus with affection For when one loves one's sex mate, one has all kinds of conflicts in relation to her: e.g., shall I marry her? shall I go to the movie she likes instead of the one I like? shall I let her get away with those names she is calling me? etc. When one does not love one's sex mate, decisions in regard to her tend to be much simpler and less conflicting.

The conflict that you talk about, between giving in to one's desires for non-affectional sex and this activity's being "unacceptable to the rest of one's personality" is a conflict that is, to the very largest extent, unnecessarily CREATED by our idiotic sex teachings, and that only to a small degree stems from the intrinsic conflict between accepting one pleasure (sex without love) and giving up another pleasure (e.g., spending the time, instead, to read a good book or court a girl who might love one). If we did not TEACH that sex without love is, to use your term, "unjustified," I am sure that MOST of the self-conflict concerning it would vanish. It would then come down to a simple decision as to whether an individual would have non-affectional sex or engage in some other activity, including striving for sex with love. This decision might well create SOME amount of intrapersonal conflict—but hardly the amount, I am sure, that you see it creating under our PRESENT sex mores.

In other words: I am hypothesizing that the more we teach people that sex without love is wicked or worthless the more conflict we CREATE in them when they have such sex relations. Under the societal code you propose, therefore, such conflict would be appreciably increased.

6. You parenthetically note, on page 2, that love without sex might well lead to an increase in adultery. Assuming that such an increase is an undesirable thing (which, of course, it has never entirely been proven to be), I would again hypothesize just the reverse: namely, that an increase in the notion that sex must be

had with love would almost certainly lead to an increase in adultery. For in most marriages, for reasons which I need not go into at the moment, very little love seems to remain after the first few years. But sex relations, for the most part, continue. If these sex relations without love were to stop, then perhaps the majority of men and women who have been married for, say, ten or more years would be driven to adultery—where it is quite probable that they would find sex with love. The notion that sex should and, indeed, must go with love is a rather romantic notion that does not square at all well with the most modern sociological concepts of marriage, since most of the "experts" in the field (including, if I may so so, myself) keep writing that marriage, in order to be lastingly happy, should probably be somewhat de-romanticized. Both love, or at least romantic love, and sex satisfaction, I would say, are often most easily maintained if the individual who seeks them changes his partners frequently. Consequently, if you INSIST that love and sex must go together, I believe that you will soon doom the continuation of millions of marriages that now go on, albeit none too ecstatically, not too miserably.

To summarize what seem to be the differences between our views: We both agree that sex with love is a good thing and that it may be better for most human beings than sex without love. You, however, believe that it is virtually ALWAYS better; I think that it USUALLY or OFTEN is. You believe that sex without love bases and drives out sex with love; I think that it usually does not and that, in Fact, non-affectional sex often leads to and serves as a learning experience for affectional sex. You think that loveless sex is almost invariably a product of immature, self-defeating behavior; I think that NOT experiencing it, when sex with love is unavailable, is a product of immature, self-defeating behavior. You believe that, in the long run, limiting oneself only to sex with love brings greater satisfaction than giving in to loveless sex; I think that it brings, in both the short and the long run, less satisfaction. We both agree that man's biological sex urges require social restrictions, but I believe that these restrictions should be minimal and you seem to believe that they should be more than minimal. You believe that having sex relations without love leads to intrinsic personality self-conflicts; I believe that most of the self-conflict that thereby arises is instilled by superfluous and happiness sabotaging teachings of society. You believe that if we sanction sex without love it would lead to an increase in adultery; I think that, in our present culture, insisting on sex with love would lead to an increase in adultery.

If my summary is correct, then it would appear that although in some ways our differences are slight-since you essentially believe that sex with love is almost always a GOOD thing and I believe it is usually, though not always, a PREFERABLE mode of behavior—in other ways our main derivative hypotheses are mutually exclusive. The irony is that in regard to a basic philosophy of human conduct, we are probably not far apart. For my own philosophy of life, which I have recently expounded in several unpublished papers on what I call rational psychotherapy, is that human beings should learn to think clearly so that they can often give up present pleasures for future gains and can act "rationally" instead of "emotionally". In fact, I spend most of my professional time teaching my clients to do just this: to accept the hard facts of reality and to give up childish, unrealistic "pleasures" for more mature, more global satisfactions You seem to believe, however that it is basically irrational for humans to have sex without love when they can almost invariably have it with love; and I think that it is basically irrational for them not to have sex without love when, for the nonce, they cannot have it with love. In the final analysis, I don't see how we can answer the questions we pose without considerable experimental data. If you can suggest a practical way to obtain that kind of data, I should be quite interested in getting a relevant study under way.

Incidentally, I shall probably use some of the contents of these letters to you as the basis of my next column in THE INDEPENDENT. I shall start the column somewhat along the following lines: "I have recently been having a most interesting correspondence with one of my professional colleagues on the subject of whether it is more desirable for men and women to have sex relations without love or to confine their sexual activities only to partners with whom they have some kind of an affectional involvement. . . ." I can easily say, instead, ". . . one of my professional colleagues, Dr. Ira L. Reiss, professor of sociology and anthropology at the College of William and Mary . . ." Please let me know whether or not you prefer to remain anonymous in this respect. I shall probably quote little or none of your views in the column, since I imagine that you would want to speak for yourself, but shall stick only to my side of the issue.

Cordially yours,

Albert Ellis

Comments by Albert Ellis on His February 8, 1957 Letter

I would endorse most of my statements in the letter today, except for the somewhat sexist language. Ira and I say, "Man can be trained to act sexually anyway at all." We would better have said, "*People* can be trained." Again, "man's biological sex urges require social restrictions" could better be restated as "men and women's biological urges."

I also say, "Very little love seems to remain after the first few years" of marriage. I could have more accurately said, "Very little romantic love seems to remain after the first few years" of marriage, as I indicate later in this letter and as I also showed in my book *The American Sexual Tragedy* (1954).

COLLEGE OF WILLIAM AND MARY
WILLIAMSBURG, VIRGINIA

February, 13, 1957
Wednesday Afternoon

Dear Dr. Ellis:

I should be busy preparing lectures but it's a lazy, sunny day out and I'd much prefer to answer your letter, so I will.

There are a few points in your last letter that I would like to comment on. First off, I am not in favor of restricting sexual behavior or any sort of behavior just for the sake of restriction or for the sake of enjoying the act more latter due to the privation. I am opposed only to that sexual behavior which causes one to lack the time or the desire for coitus with affection. Even if we agree that sex without affection will not lead one to violate other values of his, that it will not become habitual, and so forth, I think you must admit that coitus without affection takes time and if this is so then it takes time that could otherwise be spent meeting and getting to know people who could become affectionate sexual partners. Thus it seems quite clear to me that the more one devotes his time and efforts to coitus without affection. the less one will have time and effort to devote to coitus with affection. Since I feel that coitus with affection because of its psychic element is much more worthwhile than the predominately physical coitus without affection, then I accordingly feel that only in infrequent situations should one devote time to coitus without affection. It seems better on my value system that one spend the time in between affairs looking for an "affection partner" rather than devoting any considerable part of it to coitus without affection. I also do feel that because coitus without affection is relatively easy to obtain one might well take the path of least resistance and habitually engage in it, and thus it should not be engaged in too often to avoid this. But even if this is not so, even if one would still in practice as well as ideology (for it is different to say you like affectionate coitus than to actually seek and obtain it), adhere to coitus with affection, even so the time factor mentioned above seems to be a valid point. <u>Thus only in rare instances as a diversion, a "fill in", or due to an usually strong physical attraction, would I feel that coitus without affection would be justified. It is not worth the time in other cases, since the time can be better used getting to know possible affectionate partners.</u>

Secondly, you are a psychologist and I am a sociologist and the difference shows clearly in our letters. You are constantly referring to cases of individuals who are in unusual circumstances and therefore are justified in divergent behavior. You mentioned Immanuel Kant in your last letter and his addiction to coitus without affection—I am not too fond of Kant and find him most difficult to fathom and his style strikes me chaotic, but I do not give a damn what he does to satisfy his sexual desires and I suppose I would agree if it helped him contribute something significant to our cultural heritage it would certainly be justified. I fully recognize the pressures of circumstances and individual differences for example say that for a traveling bachelor coitus without affection may be the only real alternative to abstinence. So here too I would accept a divergent position. However, as a sociologist, I am primarily interested in a social code, its workability and consequences and only secondarily interested in small numbers of people who because of peculiar circumstances or personality require special attention. But to avoid controversy let me say that I am certainly aware of such deviances.

Thirdly, you seem to have assumed that I used the words "affection" and romantic love interchangeably. I did not. I meant by affection all strong feeling states from "strong affection" to love, which are based upon a great deal of personal contact, of seeing of one another in varied situations over a period of several weeks at least—I did not mean quick love at first sight feelings or feelings based on seeing each other only on Saturday night dances. Coitus with affection is to me a mature stable relation and not a concession to romantic illusions.

Fourthly, I would agree that coitus without affection be viewed as offering positive value opportunities if not abused; I agree that we should not unnecessarily make people feel guilty. But I also feel that the value of coitus without affection is so small as compared to the value of coitus with affection that we should not encourage people to sacrifice the larger value for the smaller one. When coitus without affection occurs rarely in exceptional circumstances then I say it is good and probably is a rewarding experience but since coitus with affection is worth so much more we must control the situation by channeling peoples' time and effort into coitus with affection. If in doing this we make those people who devote much of their time and effort to coitus without affection feel guilty, then I do not feel sorry for I believe that this is necessary to insure an emphasis on what is more important, coitus with affection. Guilt feelings that prevent or tend to prevent people from moving in a direction that is wrong are surely not to be eliminated and this is the kind of situation that exists here, I believe.

Fifthly, I think I may make myself clearer if I use a few analogies. I see nothing wrong with eating candy, listening to popular music or even watching the quiz shows on TV. All of these things are of value but (in my mind) of very small value. Now, when eating candy starts to interfere with eating nourishing food, when listening to popular music starts to interfere with listening to Jazz or Classical music, when watching Quiz Shows interferes with watching Hamlet or

Mayerling, then and only then, I say this is a bad situation and something should be done. If we must call such behavior bad and make people who perform it feel guilty, and if this will prevent such action, then I say let us do it. These other things are of so much more importance that it seems best to devote all most all of one's time and effort to pursuing them and to only infrequently devote one's most limited and precious time to candy, popular music and quiz shows. To do otherwise would be not to use your time to the best advantage. So I feel it is with coitus, to devote your time in any significant amount to coitus without affection is to detract from the time you devote to a much more meaningful mode of coitus, coitus with affection.

Sixthly, and lastly, let me say that I feel that we should _not_ teach that coitus without affection is unjustifiable or bad, or such, but that it is of relatively slight value as compared to coitus with affection and should be accordingly restricted. I feel that this is the minimal, necessary, restrictions for a sexual code in America today. I think our differences can be summed up in one sentence: I valued coitus with affection more than you do and value coitus without affection less than you do. You seem to feel that there is relatively little difference and are not particularly concerned whether people engage in one or the other. However, you also state that coitus with affection is better. This puzzels me somewhat for if you do feel this way, why are you opposed to restricting coitus without affection so as to guarantee more time be devoted to coitus with affection?

I feel the preceeding two pages are a good statement of my views and I think they summarize our differences. I will be most interested to hear your comments.

Please do send me a copy of your article for the Independent. I think it would be better not to mention me by name and I also agree that it would be best to state mainly your own views for I would want to speak for myself, as you supposed.

Sincerely,

ALBERT ELLIS, Ph. D.
PARC VENDOME
333 WEST 56TH STREET
NEW YORK 19, N. Y.

February 22, 1957

Dear Dr. Reiss:

   Your letters are always thought-provoking; and although I, like you, certainly have much else to do in life than interminably correspond, I enjoy answering the points you raise. As usual, let me take the items in your last letter seriatim.
   1. I can well agree with you <u>in theory</u> that time spent in getting sex partners on a non-affectionate basis may well interfere with time spent in having, or arranging for, affectional love affairs. <u>In practice</u>, however, I think you are being somewhat unrealistic. Normally, a boy dates a girl to see if she will be his "ideal" mate, whom he can love and marry. In the vast majority of instances, however, he quickly finds that she isn't. What is he to do, then: immediately leave her at, say, 9 P.M. on a Saturday date, and go off to find a more suitable girl, whom perhaps he can love? Naturally not. We expect, at the very least, that out of human politeness he will try to afford her a good evening's entertainment. And if, in the course of this entertainment, he can have mutually satisfying and voluntarily entered sex relations with the girl, why the devil should he not? This kind of situation, it seems to me, is, far from being a rare occurrence, the <u>usual</u> par for the dating course.
   Now let's be still more realistic. A boy, having met a girl in class, or at a dance, or through an introduction, knows perfectly well that she's not amatively or maritally for him—or at least strongly suspects that she isn't. But, at the moment, he knows no other girl that is. What, again, is he to do with his dating nights—just go out to look for, which would literally mean in most instances to pick up, a girl who MIGHT (but who, most probably, would not) be someone he can love? Or, faute de mieux, is he hard-headedly to date the girl he knows he probably won't love and try to have sex relations with her?
   Now let's be just a little more realistic. A boy, knowing that he doesn't love any girl at present, but also knowing that, on a statistical basis, he could eventually find such a girl, decides that he will look for love with sex rather than sex without love. Fine. So he looks and looks but finds that it takes him quite a while, especially with the limited time and number of possible love mates he may have at his immediate disposal, to locate the girl of his dreams. In the meantime, his sex desires are hardly non-existent. So, until he finds his true love, he has a

choice between (a) masturbating and (b) having sex relations with a non-true love. What would you suggest he do?

Let's take another—and still very common—case. A boy finally, after much seeking finds the girl he loves and attempts to have sex-love relations with her. But, for various good or bad reasons, she refuses. She either doesn't, as yet, sufficiently love him; or she does, but still doesn't want to respond sexually. He still feels he loves her, and wants to continue the relationship on a socalled platonic level, until she comes around to including sex in it; and he also happens to know another girl whom he doesn't love, and who probably doesn't care for him either, but who is sexually freer than his dearly beloved. What would you suggest, again, that he do—remain pure because his beloved, perhaps for several years to come, sets up a distinct sex barrier?

Another practical problem. Suppose that a boy and a girl really stick to the letter of your rule, and have sex relations without love "only in rare instances". Do you really think that they would ever get any amount of sex experience so that, in this culture, they would fulfill themselves in a sexual way when they finally did meet their soulmates? Even socalled normal young people, as I found in a study I did some time ago, and as Hamilton and others have also found, love only half a dozen times or so before they settle down in marriage. And perhaps half or three-quarters of these loves are one-sided or for one reason or another fail to lead to sexual fruition. This means that in the most highly sexualized parts of their lives (say, between the ages of 15 and 30), they are going to be lucky to have three or four years of potential steady sex satisfaction—if they stick to your criteria. Do you think, from a practical or any other standpoint, that is a good thing?

2. I quite agree with you that we should not emphasize individual cases and should seek rules for the general populace. But my feeling is that your rule would lead to considerable sexual suffering for MOST of the population. I am not only fighting for the Immanuel Kants of the world (he, incidentally, was addicted to NO coitus rather than coitus without affection) but for what I believe is the majority of both "normal" and abnormal human beings. Maladjusted people have an even more difficult time becoming affectionate or falling in love than so called adjusted ones; and I am especially sensitive to their problems being, as you point out, a psychologist. But the AVERAGE male and female, I believe, would distinctly suffer from your sex-should-only-rarely-be-had-without-affection rule.

3. I'll accept the view that affection need not be romantic love. But even what you call "strong affection" is not an easy state to attain between young people. And although I still think it a desirable state as an accompaniment to sex, I fail to see it as a necessary one.

4. Our basic disagreement seems to be that you consider "the value of coitus without affection is so small as compared to the value of coitus with affection." I

think that coitus without affection has a very strong, positive value in, of, by, and for itself—especially for young people; and that the fact that coitus with affection may have a still stronger, more positive value should not be used as an argument to denigrate the value of sex without love. I personally feel that a fine steak or chunk of roast pork is distinctly more nutritious and satisfying than a box of candy; but I would hardly go around trying to turn people against eating candy. If <u>they</u> think, for some benighted reason, that the candy is better than the steak, that is <u>their</u> value, and I disagree with but respect it. Similarly, if millions of people think that sex without love is as good as or better than sex with love, I again disagree with but respect their evaluation. Maybe they <u>would</u> be better off if they gave up loveless sex for sex with affection; but, especially in our current anti-sexual and anti-loving society, I doubt it.

5. Interestingly enough, I wrote the previous section, with the steak and candy analogy, before I read your fifth point, which uses a similar analogy. In reading your use of the analogy, I feel even more strongly attached to mine. For I quite disagree with you: even though I would prefer classical to popular music, and Hamlet to quiz shows, I would say that as long as human beings are constructed and raised any way similar to the way in which they are constructed and raised today, we should (a) by all means provide popular music and quiz shows for them and (b) under no circumstances make them guilty about enjoying these kind of entertainments. Personally, I would like to see the kind of a world where humans would be more intelligent, creative, and cultured than they now are. I believe that such a world will never exist until, along with a pronounced change in our education and our socio-economic system, we also resort to some kind of eugenic selection. If such a brave new world as I would like to see exist actually did, I do not doubt that there would be in it relatively more sex with and less sex without affection than now is true. But this kind of a world does NOT exist; and for, say, the next several centuries at least, is not likely to exist. Why, then, should we create needless guilt and despair among the kind of people in the kind of world that DOES now exist?

Moreover: though I prefer classical to popular music and Hamlet to quiz shows, I would hate to see a world in which there was NO popular music or quiz shows. Even highly intelligent, educated, and cultured individuals, I feel, will SOMETIMES want to enjoy popular music and quiz shows. Similarly, even individuals who LARGELY enjoy sex with affection will, I feel, SOMETIMES thoroughly, even ecstatically, enjoy sex without affection. And why should they not?

6. In the light of what I have just said, I see no contradiction in my feeling (a) that coitus with affection is USUALLY better for MOST intelligent people for MOST of their sex involvements and my feeling (b) that coitus without affection

should not be restricted so as to guarantee more time that can be devoted to coitus with affection. My position in this respect parallels the position I took in a letter I wrote earlier this evening to a psychologist who asked me what was my attitude toward forced conciliation when people come to court for marital difficulties. I replied that (a) I definitely thought, on the basis of my experience, that individuals who are forced into psychotherapy or counseling can often, even though it is at first against their will, come to benefit from this counseling and thus save their marriages; but that (b) I would not, under any circumstances, want to have them forced, either by direct coercion or making them terribly guilty, into resorting to conciliation. I am more—perhaps idealistically—attached to the notion of the sacredness of human individuality and freedom than I am to the notion of forcing people to be "better" than they presumably are.

Similarly, granting that people, or SOME people, might benefit by being forced, through guilt or other coercion, to stick to affectional rather than non-affectional sex relations, I would not want to see such "benefits" brought about. I think that the price is far too much to pay—especially since I am utterly convinced, after long clinical experience, that virtually all modern neurosis is basically caused by our teaching people to be guilty or self-blaming. I am reminded, in this connection, of a recent talk by Donal J. MacNamara, an authority on criminology, who pointed out that in dictatorships they invariably have much less gangsterism and criminality, in the usual sense, than we have in our kind of a democracy. The question he raised was: Is the social cost of such "benefits" too high? I think it is.

I agree that our differences on this matter of affectional coitus seem to be fairly well summarized by now. We tend to have distinctly different values in relation to the worthwhileness of sex with and without affection. We also seem to differ significantly as to whether it is most desirable to try to change human beings radically, granting that they engage in some relatively undesirable conduct, or whether it is best to let them, over a long period of time, gradually work out their relatively poor behavior for themselves. Although I used to be a social revolutionist, in my latter days I seem to be turning more and more in favor of more evolutionary change processes. Maybe it's old age creeping up on me.

I haven't yet written my INDEPENDENT article on the topic of our discussion; but if I do so, I shall certainly send you a copy.

Sincerely,

Albert Ellis

Comments by Albert Ellis on His February 22, 1957 Letter

I now go along with the main points I made in this 1957 letter. However, the point I made about our possibly improving ourselves by a "pronounced change in our educational and socio-economic system" and "we also resort to some kind of eugenic selection" has its severe problems. Who is to do the "eugenic selection" and how, if feasible, is it to be done—that would be quite a problem!

My point that "I am more—perhaps idealistically—attached to the notion of the sacredness of human individuality and freedom" is put too strongly. I and REBT teach today that *nothing*, including human individuality and freedom, is *sacred*—only, by choice, important. Human individuality and social responsibility—both/and, not either/or—are the way to go!

ALBERT ELLIS, Ph. D.
PARC VENDOME
333 WEST 56TH STREET
NEW YORK 19, N. Y.

March 1, 1957

Dr. Ira L. Reiss
Dept. of Sociology & Anthropology
College of William & Mary
Williamsburg, Va.

Dear Dr. Reiss:

As I promised you in my last letter, I am enclosing herewith a copy of the article I just sent to THE INDEPENDENT for publication in its next issue. As you will see, it sticks fairly closely to the contents of my first letter to you.

I may use material from my other letters for another article in this series. If so, I shall send you a copy.

Cordially,

Albert Ellis

The Justification of Sex Without Love

Albert Ellis, Ph. D.

A scientific colleague of mine, who holds a professorial post in the department of sociology and anthropology at one of our leading universities, recently asked me about my stand on the question of human beings having sex relations without love. Although I have taken something of a position on this issue in my book, <u>The American Sexual Tragedy</u>, I have never quite considered the problem in sufficient detail. So here goes.

In general, I feel that affectional, as against non-affectional, sex relations are <u>desirable</u> but not <u>necessary</u>. It is usually desirable that an association between coitus and affection exist—particularly in marriage, because it is often difficult for two individuals to keep finely tuned to each other over a period of years, and if there is not a good deal of love between them, one may tend to feel sexually imposed upon by the other.

The fact, however, that the coexistence of sex and love may be desirable does not, to my mind, make it necessary. My reasons for this view are several:

1. Many individuals—including, even, many married couples—<u>do</u> find great satisfaction in having sex relations without love. I do not consider it fair to label these individuals as criminal just because they may be in the minority. Moreover, even if they are in the minority (as may well <u>not</u> be the case), I am sure that they number literally millions of men and women. If so, they constitute a sizeable subgroup of humans whose rights to sex satisfaction should be fully acknowledged and protected.

2. Even if we consider the supposed majority of individuals who find greater satisfaction in sex-love than in sex-sans-love relations, it is doubtful if all or most of them do so for <u>all</u> their lives. During much of their existence, especially their younger years, these people tend to find sex-without-love quite satisfying, and even to prefer it to affectional sex. When they become older, and their sex drives tend to wane, they may well emphasize coitus with rather than without affection. But why should we condemn them <u>while</u> they still prefer sex to sex-love affairs?

3. Many individuals, especially females in our culture, who say that they only enjoy sex when it is accompanied by affection are actually being unthinkingly conformist and unconsciously hypocritical. If they were able to contemplate themselves objectively, and had the courage of their inner convictions, they would find sex without love eminently gratifying. This is not to say that they would <u>only</u> enjoy non-affectional coitus, nor that they would always find it <u>more</u> satisfying than affectional sex. But, in the depths of their psyche and soma, they would deem

sex without love pleasurable _too_. And why should they not? And why should we, by our puritanical know-nothingness, force these individuals to drive a considerable portion of their sex feelings and potential satisfactions underground?

If, in other words, we view sexuo-amative relations as desirable rather than necessary, we sanction the innermost thoughts and drives of many of our fellowmen and fellowwomen to have sex _and_ sex-love relations. If we take the opposing view, we hardly destroy these innermost thoughts and drives, but frequently tend to intensify them while denying them open and honest outlet. This, as Freud pointed out, is one of the main (though by no means the only) source of rampant neurosis.

4. I firmly believe that sex is a biological, as well as a social, drive, and that in its biological phases it is essentially non-affectional. If this is so, then we can expect that, however we try to civilize the sex drives—and-civilize them to _some_ degree we certainly must—there will always be an underlying tendency for them to escape from our society-inculcated shackles and to be still partly felt in the raw. When so felt, when our biosocial sex urges lead us to desire and enjoy sex without (as well as with) love, I do not see why we should make their experiencers feel needlessly guilty.

5. Many individuals—many millions in our society, I am afraid—have little or no capacity for affection or love. The majority of these individuals, perhaps, are emotionally disturbed, and should preferably be helped to increase their affectional propensities But a large number are not particularly disturbed, and instead are neurologically or cerebrally deficient. Mentally deficient persons, for example, as well as many dull normals (who, together, include several million citizens of our nation) are notoriously shallow in their feelings, and probably intrinsically so.

Since these kinds of individuals—both the neurotic and the organically deficient—are for the most part, in our day and age, _not_ going to be properly treated and _not_ going to overcome their deficiencies, and since most of them definitely _do_ have sex desires, I again see no point in making them guilty when they have non-loving sex relations. Surely these unfortunate individuals are sufficiently handicapped by their disturbances or impairments without our adding to their woes by anathematizing them when they manage to achieve some non-amative sexual release.

6. Under some circumstances—though these, I admit, may be rare—some people find more satisfaction in non-loving coitus even though, under other circumstances, these _same_ people may find more satisfaction in sex-love affairs. Thus, the man who _normally_ enjoys being with his girlfriend because he loves as well as is sexually attracted to her, may occasionally find immense satisfaction in being with another girl with whom he has distinctly non-loving relations. Granting that this may be (or is it?) unusual, I do not see why it should be condemnable.

7. *If many people get along excellently and most cooperatively with business partners, employees, professors, laboratory associates, acquaintances, and even spouses for whom they have little or no love or affection, but with whom they have certain specific things in common, I do not see why there cannot be individuals who get along excellently and most co-operatively with sex mates with whom they may have little else in common.*

I personally can easily see the tragic plight of a man who spends much time with a girl with whom he has nothing in common but sex: since I believe that life is too short to be well consumed in relatively one-track or intellectually low-level pursuits. I would also think it rather unrewarding for a girl to spend much time with a male with whom she had mutually satisfying sex, friendship, and cultural interests but no love involvement. This is because I would like to see people, in their 70-odd years of life, have maximum rather than minimum satisfactions with individuals of the other sex with whom they spend considerable time.

I can easily see, however, even the most intelligent and highly cultured individuals spending a <u>little</u> time with members of the other sex with whom they have common sex and cultural but no real love interests. And feel that, for the time expended in this manner, their lives may be immeasurably enriched.

Moreover, when I encounter friends or psychotherapy clients who become enamored and spend considerable time and effort thinking about and being with a member of the other sex with whom they are largely sexually obsessed, and for whom they have little or no love, I mainly view these sexual infatuations as one of the penalties of their being human. For humans are the kind of animals who are easily disposed to this type of behavior.

I believe that one of the distinct inconveniences or tragedies of human sexuality is that it endows us, and perhaps particularly the males among us, with a propensity to become exceptionally involved and infatuated with members of the other sex whom, had we no sex urges, we would hardly notice. That is too bad; and it might well be a better world if it were otherwise. But it is <u>not</u> otherwise, and I think it is silly and pernicious for us to condemn ourselves because we are the way that we are in this respect. We had better <u>accept</u> our biosocial tendencies, or our fallible humanity —instead of constantly blaming ourselves and futilely trying to change certain of its relatively harmless, though still somewhat tragic, aspects.

For reasons such as these, I feel that although it is usually—if not always— <u>desirable</u> for human beings to have sex relations with those they love rather than with those they do not love, it is by no means <u>necessary</u> that they do so. And when we teach that it <u>is</u> necessary, we only needlessly condemn millions of our citizens to self-blame and atonement.

Comments by Albert Ellis on His Attachment to His March 1, 1957 Letter

The article "The Justification of Sex without Love," which was published in the *Independent* and later in my best-selling 1958 book, *Sex without Guilt*, restates the points I made to Ira in my previous letters. I still hold these main views—yes, over forty years later!

I again disagree, however, with my statement "sex is a biological, as well as a social, drive, and in its biological phases it is essentially non affectional." No, this sentence is inaccurate—as I should have seen if I consistently followed my own views that I presented in August 1956. In its biological phases, sex is significantly behavioral *and* emotional, physiological and affectional.

I also state that people have "70-odd years of life." Today, this has been raised to eighty-odd and it is still climbing.

COLLEGE OF WILLIAM AND MARY
WILLIAMSBURG, VIRGINIA

March 4, 1957

Dear Dr. Ellis:

 I still feel that time devoted to sex without affection interferes with sex with affection in theory <u>and</u> in practice. You give the example of a boy and girl who shortly realize that they cannot come to really like or love each other—it is early in the evening—what are they to do? Of course, they see the evening out but having intercourse is not such an innocuous event as you assert. Intercourse is a pleasurable experience and thus if one decides that even though he doesn't care for this girl he'll have coitus with her anyhow, it may well be and I firmly believe it often is, that this act of intercourse does not end the relationship but rather may prolong it. Thus whereas without having intercourse the boy and girl would just stop seeing each other, by having intercourse they may continue to see each for the sake of pleasure and refrain from dating other people whom they could possible get to love. Thus coitus without affection does interfere in practice as well as in theory for it tends to occupy one and prevent one from finding love-partners.

 You also say there is a practical difficulty in men having to fill in the time until they meet they future mates. Of course, there is, but I have always defined coitus with affection as including all strong and stable affection states not Just those rare instances of marriage-type love. Thus the men concerned (and the women also) can fill in their "spare time" by looking for a relationship that as a minimum involves strong fondness. This is not a choice between masturbation and non-affectional intercourse. From a practical standpoint I would say that the vast majority of people could meet such an affection-partner at least once or twice a year and carry an a prolonged affair with them. Thus the waiting periods are relatively short. Of course, if you are only interested in reaching orgasms in every spare moment of life then this is not workable but such activity would lead to establishing habits and relations which would take away from the time one had available for sex with affection, and thus you would agree that it was not advisable

 Your example of the boy whose lover refused intercourse again shows your feeling that coitus has practically no consequences of note. If you want to be practical in this case you surely cannot avoid the jealousy feelings that would be aroused in most women if their boyfriends cheated on them. In fact, if you want to be practical the seeking of sex without affection seems to have many drawbacks—you may become too eager to have such sex and since most American females do not accept this standard, your acceptance of it may just lead to greater frustration. Also, you may find yourself with a jealous lover because you are satisfying yourself elsewhere with sex-without-affection, which would not occur if you did not accept such sex. So from a practical standpoint I think sex without affection has many pitfalls in America of today. I have always during our discussions thought of it in terms of what would be if all accepted it. From a practical standpoint it seems most unwise since it seems so capable of generating conflicts, like the above.

Your statements about the suffering (due to privation) that the average person would undergo are based on restricting sex to marriage type love, as I've said in previous letters, "With-Affection" includes to me all strong and stable affectionate states. Thus this suffering would be diminished. Further, the amount of suffering is not a thing which must vary with the amount of restraint, but rather with the amount of restraint *felt*. In our sex-tease culture this restraint that is felt is quite high but if people were brought up to accept sex-with-affection the felt restraint might be negligible. From a practical standpoint there is only a very remote possibility of people in america accepting sex-without affection while there is a very real possibility of people accepting sex-with-affection. Thus the practical course of action it seems is to get rid of abstinence and take sex-with-affection, rather than to try and get without-affection-sex and have people refuse to go that far and then be stuck with abstinence.

I, of course, am not saying that people should not be allowed to follow whatever path of action they feel is right. I am saying that society should set up its ideals so as to encourage people to make the most of the time they have in this world. I think a society should afford a wide variety of outlets, including sex without affection, candy and quiz shows (I've tried and liked all three at various times). But I think we must build a hierarchy of values, I think, to be organized a society must value some things more than others, or else social preference and individual choice become seriously confused. To say that one must always do one thing is to my mind bad, to force one at gun point to do something, to my mind is bad, but to tell someone that A is worth more than B, and that he can do both, but to keep in mind their relative worth, this is a good and necessary thing, to my mind. If as a result of this, people feel some qualms when they stress the lower value to the exclusion of the higher, then this is the price one must pay in order to achieve the higher value. We cannot avoid all pain, life has pain in it and all we can do is avoid unnecessary pain. I fail to see how pain which keeps people seeking what they agree is good, is not necessary. This is not dictatorial, unless you think that all social norms are dictatorial. For all societies have rules concerning what is valued more than what else. You cannot do away with all these rules for then we would not know what to expect of one another and you would have created more chaos and conflict then before. To minimize restrictions, to give man a wide range of outlets, are wonderful goals, but to try and do away with all restrictions and with all comparisons is to destroy the thing you are trying to save. It is only in an organized society that freedom is meaningful, without organization we would never know what to expect from others—a pat on the shoulder or a knife between them. A life with a minimum of guilt and a maximum of freedom is to be desired, but a life without guilt or values, is a life without evaluations and that is unknown to mankind so far, or in the realm of the possible. Our animal ancestors and contemporaries are more equipped than we are for that type of life, they have poorer memories and cannot think of the future as well. We can do these things and thus we will inevitably set our sights and evaluate our goals and examine our past.

Let me summarize my feelings, once more. I can see that we differ somewhat in our evaluation of with- and without-affection-sex; I can also see that we differ somewhat concerning the role of social norms in regulating behavior. These differences do not bother me as they are genuine differences in values concerning sex and freedom. But I do feel that despite these differences you would still agree that you would place some limits on coitus without affection—less than I would since you value it more and value freedom more, but limits nonetheless since you say you value sex with affection more than sex without affection. It is on this point that I feel you are inconsistent. Your position here is more consistent with a belief that both types of sex are of equal value.

I would agree that there are many men who have abundant amounts of sex without affection and still say they prefer sex with affection. But these very same men devote the vast majority of their time to sex without affection and to meeting new without affection partners. This indicates clearly that their preference is not very great and/or is more on the intellectual then the emotional level. Similarly there probably are many people who gorge themselves with candy all the time but who would admit that other foods are much better. Here too the emotional attachment to the candy is quite strong. It seems to me that unless you place restrictions on sex without affection you will end up with these type of people. They are produced in situations like you mentioned, where they have coitus because there's nothing else to do and keep it up because they like it and when they get tired look for another quick thrill which is so much easier to obtain then the affectionate sex. The product is an individual who is emotionally enslaved to a standard he himself thinks is not as good as another one. Is this the freedom you want? Is this the product you want? Can you deny that this is a very common result from the attitude that no restrictions should be placed on the nature of coitus? Is this not what happens when you fail to emphasize the relative *importance of values? Would not the man who restrained himself but who thereby followed in theory and practice what he thought was a more worthwhile form of behavior, would not this man have more freedom? Is it not reasonable to assume that the more one has coitus without affection, the more he will become habituated to it and occupy his time with this type of coitus and the more likely he will be to repeat his previous behaviors and get involved in relations that take time and lack affection? If so, then must not we agree that in practice as well as theory, sex without affection does detract from the time one can devote to sex with affection and thus, although it is allowable and satisfying, it should be kept under control and minimized?*

COLLEGE OF WILLIAM AND MARY
WILLIAMSBURG, VIRGINIA

March 5, 1957

333 West 56th Street
New York 19, New York

Dear Dr. Ellis:

    Our letters crossed in the mail—I wrote you an answer to your letter of February 22nd yesterday and received your <u>Independent</u> article today.

    We've covered most all the points in your article in our letters. As always we seem to agree that sex with affection is more desirable than sex without affection and that there are exceptions to this rule, for sex without affection in certain situations does have value. But as usual here our agreement ceases for you go on to say that people who prefer sex without affection for their whole life or for part of it should not be inhibited by our stressing the value of sex with affection. It seems to me that by being so free you stress the value of sex without affection so much that it tends to vitiate your original statement about the greater desirability of sex with affection.

    However, at the end of your article you state that you recommend spending a <u>little</u> time on sex without affection. With this statement I would certainly agree—I have never held that sex with affection is always necessary but rather have viewed it as preferable <u>under most conditions</u>. Nevertheless, I am still bothered by what seems to me an inconsistency in your saying that young men, etc. can have all the sex without affection they want and then ending up by saying you recommend just a <u>little</u> of sex without affection and a large doseage of sex with affection. If having a <u>little</u> of coitus without affection is the way to maximize satisfactions, as you say, then would you not say that most people (excluding morons) should follow that example and that even if young men desire and enjoy concentrating on sex without affection it would be well if we could induce them to enjoy sex with affection?

    I'll close here because I think we've covered most all the other points. I did enjoy reading your article and believe it will probably start many people thinking on this topic—which should be a good thing and may clear the sexual air a little.

Sincerely,

Ira L. Reiss
Assistant Professor
Department of Sociology
and Anthropology

ALBERT ELLIS, PH. D.
PARC VENDOME
333 WEST 56TH STREET
NEW YORK 19, N. Y.

March 9, 1957

Dear Dr. Reiss:

As usual, I shall answer the points in your last two letters seriatim:
1. You are quite right that if a boy finds sex satisfaction with a girl, and vice versa, they may continue to have intercourse in lieu of looking for more loving partners. But they also may <u>not</u>. I, personally, would teach them to continue having intercourse <u>and</u> look for still better, or more loving, partners. You apparently feel that the mere fact of their having intercourse discourages them from looking. I thus seem to have more faith in their ultimately finding love, in spite of satisfactory non-loving sex relations, than you have. I have to admit, however, that in a puritanical society like ours, where satisfactory sex relations are made scarce, an individual's finding such relations may even prejudice him against looking for sex-love relations. In which case, say I: let us get rid of the puritanism, make the finding of non-affectional sex relations very easy and hence relatively valueless, and thus promote sex-love affairs.
2. I agree that making sex-love affairs, rather than just marriage-intending affairs, desirable and free you reduce the waiting period for the non-married. But the fact unfortunately remains that, in our present culture, females in particularly feel compelled to marry, and to marry in time to raise a family, etc. Consequently, they largely tend to limit their sex-love affairs to males whom they believe may be good marriage partners. Although more liberal young people in our culture may have one or two sex-love affairs a year, most females get quite disturbed after having several such affairs and begin to worry about marrying. The more conservative females—still, alas, the majority—are far from having one or two affairs a year. To have your system work, therefore, you would have to arrange for a drastic change in our sex mores—which, of course, I hardly oppose.
3. I agree that sex-without-affection has its grave difficulties in OUR society. But, as I just noted above, so does sex-with-affection. In fact, I cannot see how ANY sex code can lead to anything but continual disasters in this culture, since we have a basic anti-sexual foundation; and this, as you point out, leads to jealousy, double standards, etc. The only solution that I can see is a quite DIFFERENT, non-puritanical base to begin with. Such a base, I believe, would make both your proposals and mine more realistic. Without it, we are bound to have sex chaos.

4. I agree that, from a practical standpoint, Americans can more easily be sold on sex-with-affection than sex without. At least n theory: since, in real practice, sex-without-affection has been doing very nicely for decades. But, by the same token, I am sure that, practically speaking, Americans can be sold on a fairly rational and sane code of personal ethics by calling this code "religion" than by labeling it, as it really is, non-religions or a-religious (since religion actually consists of faith unfounded on fact and has nothing intrinsic to do with ethics). For my part, however, I would rather see ethics called ethics, and not confused with religion; and the mere fact that it can be "sold" better under the name of "religion" is not, for me, a good reason for doing this selling.

5. I heartily agree that we cannot do away with all rules; and that to establish desirable standards of conduct, but not impose these on anyone, is fine. I would be distinctly in agreement with setting up a Confucian-like set of precepts, one of which might read: "Sexual congres is good in, of, by, and for itself, and should under no circumstances be disparaged. However, most human beings who have sufficient sex experience and who gain a good degree of intellectual-emotional maturity find that affectional sex relations are more rewarding than non-affectional relations and devote more of their time to the latter than to the former."

6. Contrary to yours, my position is that sex with love is (for most, though not all, intelligent people) more desirable than sex without love WHEN YOU DO NOT PLACE RESTRICTIONS ON NON-AFFECTIONAL SEX. As soon as you begin placing restrictions on sex, I contend, sex without love takes on an exaggerated significance and need. Similarly, I contend that life in which very little time is devoted to studying pornography becomes more desirable for most people WHEN YOU DO NOT PLACE RESTRICTIONS ON PORNOGRAPHY. As soon as unnecessary restrictions are placed on sex, human thought, feeling, and activity becomes distorted in relation to it, and it becomes over-emphasized rather than underemphasized.

I feel, therefore, that your position, not mine, is inconsistent. I say that sex without love is good but not preferable for most intelligent people; and I want to set up conditions of sex freedom that will allow this preference to come to the fore. You, though in a liberal manner, want to maintain many of the current sex restrictions that, in practice if not in theory, will lead to an over-emphasis on sex without love.

7. People normally engorge themselves with candy when it is not usually available, or when it is available but they are guilty about eating it. The classic technique for combatting this tendency is for a candy-store owner, when he hires a new teen-age employee, to give this employee the full run of the store: in which case, after a short while, the employee usually eats relatively little candy. By banning the employee's eating the candy, however, or teaching him that it is most

undesirable if he eats it, his appetite for candy will, perversely, be increased, and he may well continue to seek what you call "another quick thrill"—meaning, really, the thrill of doing what is banned.

8. At bottom, one of our basic differences seems to be that you think that if an immediate, minor pleasure is available, people will always prefer it to a future, more intense gain—and that therefore the immediate pleasure should be morally discouraged. I believe that, WHEN BLAME, GUILT, MORAL DISCOURAGEMENT, etc., ARE AVOIDED, human beings, especially as they grow older and wiser, learn for themselves that many future gains are worth the sacrifice of immediate pleasures. A philosophy of blame and guilt, however, effectively stops people from thinking rationally. They excoriate themselves (and their fellows) so severely for their PAST and PRESENT "sins" that they cannot possibly have the calm presence of mind and ability to observe and think straight that is necessary for their changing their FUTURE behavior.

Although, therefore, I can see your interest in "moral values" in terms of setting up desirable codes of conduct for humans to strive for, I am afraid that, unless you are exceptionally careful (as I have never seen code-setters be), you will mainly succeed in making people guilty about their past and present behavior rather than interested in achieving maximum future satisfaction. In this way, you will unwittingly defeat your own ends. That is why, as a psychologist, I stress to my patients (a) the values of self-discipline, thinking-before-acting, self-interest, etc., while concomitantly stressing (b) the necessity of not beating themselves over the head when, for the nonce, they do NOT achieve these goals. They must never, I insist, blame themselves for PAST behavior; but always ask themselves how they can do better in the FUTURE—and then practice and work like hell to improve (i.e., to become more self-interested, more loving, less blaming).

In the final analysis, then, I do not think that our goals for human beings differ so widely as they might at first seem, but perhaps our means of arriving at them. I feel that through having maximum theoretical freedom the average intelligent individual will eventually come to adopt a moderate (not too much and not too little) amount of self-discipline for his own (and, indirectly, the social) good. You seem to feel that through having a moderate degree of Freedom, the individual best achieves self-discipline. We both agree that there must be some practical restrictions and that these should not be laid on too thick.

Stated differently: although we "both agree that in practice as well as theory, sex without affection does detract from the time one can devote to sex with affection and thus, although it is allowable and satisfying, it should be kept under control," I feel that this control should be entirely individual, and should come largely from an individual's experience (including his hearing all kinds of views of others in this connection). You seem to feel that this control should be helped *considerably by propagandizing the individual with what I would think are largely one-sided philosophies (since, although I may agree with you that sex with love is ideally better than sex without love, there are surely some people who would not). Secondly: you think that sex without affection "should be kept under control and minimized" and seem to mean by "minimized" discouraged and reduced to an absolute minimum. I would only accept the word "minimized" if we mean by it placed in a <u>relatively</u> minor position in life. That is: I want sex without <u>as well as</u> with love for most normal human beings, but more of the latter than the former, at least for mature adults. You seem to want, at bottom, virtually no sex without love, on the grounds that it will drive out sex with love. Here we do not agree.*

9. In your second letter, you point out a possible inconsistency, again, in my views, but I can't quite see this inconsistency. I feel that (a) most young men should have some or a little sex without affection; (b) that, eventually, they should concentrate on sex with affection; and (c) that the best way to see that they have both (a) and (b) is not to propagandize them mightily in favor of enjoying sex with affection but to give them maximum, guiltless lee-way in this respect, or at most to give them mild Confucian-like precepts, and let them go on from there. If a young man asks me for my beliefs in this respect, I would say to him: "From my long and varied experience in the field, after trying sex with and without affection on a good many different occasions, I have found both of them good but the former distinctly better. I would therefore advise you to try both, give each a real good try, and find out for yourself which is more satisfying for you. I think you will come out of your experiences with much the same feelings as I now have. But try it and see." This, it seems to me, is somewhat less than is implied in your phrase "induce them to enjoy sex with affection."

I still feel that our means are more at variance than our ends; and it would certainly prove most interesting if we could ever set up an experiment to test our hypotheses.

Sincerely,

Albert Ellis

Comments by Albert Ellis on His March 9, 1957 Letter

I still largely agree with my sentiments as expressed in this letter to Ira. However, I would disagree with my statement "religion actually consists of faith unfounded on fact and has nothing intrinsic to do with ethics." No, there are many varieties of religious experience, as William James said, and most of them have much to do with ethics. However, atheists and agnostics also have ethical principles, so religion and ethics are not exactly the same thing.

When I say in this letter to Ira that people "must never, I insist, blame themselves for their past behavior," I am unclearly stating my and Rational Emotive Behavior Therapy position that it would be better for people to negatively evaluate or blame their past (and their present) foolish and immoral *behavior*, but *not* damn their *self* or *totality* for this behavior.

When I say, in this letter, that people's sex control "should be entirely individual and should come from an individual's experience (including his hearing all kinds of views of others in this connection)," I was inaccurate. Sex control, like other forms of self-control, is formulated in a social group in which the individual *chooses* to reside. Therefore, it cannot be *entirely* individual and has crucial social aspects.

March 18, 1957
Evening

Dear Dr. Ellis:

I enjoyed reading you letter of March 9th and think perhaps we're not so far apart after all.

You mentioned that if our puritantical attitudes towards coitus changed and non-affectionate coitus were made easy to obtain then it would tend to decrease in value and interfere less with affectionate coitus. This seems quite reasonable and I would agree with -you on this. You seem to accept my proposal of affectionate coitus which would allow people to have one or two affectionate affairs a year, but feel that people will not accept this. I would certainly agree that most Americans have not come about to this point of view but my own research and that of Burgess and Wallin, Kinsey, Terman, etc., seems to indicate that sex with affection have significantly increased in this century. Thus without getting into an argument on how soon or how probable such a change is likely to occur, it does seem that there are movements in this direction.

I think I see more what you mean when you say you do not want restrictions placed on coitus without affection. I think our differences here are merely semantic in most part. You mean you don't want people to think of any type of coitus, without affection included, as bad, evil, or forbidden. You want this in order to avoid overemphasizing any form of coitus or developing various complexes about coitus. With this I would whole heartedly agree. All I have ever been arguing for is that you teach people that although no form of coitus is bad, per se, there are forms of coitus that are much better than other forms. Once you say this you take the risk of people feeling bad because they are emphasizing a lower form and avoiding a higher form of coitus. But we must say that we feel some things are better than others—surely we cannot live together without this much organization. We can do all we can to avoid psychological difficulties but we must have a hierarchy of values. I think you must have felt that I was proposing to continue to teach that coitus without affection is bad or evil. If you check my letters I have several times stated that this is not my view, that all I want to be taught is that coitus with affection is better. I propose no other restrictions besides the belief that one type of coitus is better than another. I do not want to create or maintain any forbidden fruits ideas which will overemphasize the value of coitus without affection. I merely want to point out that there is a danger of people becoming habituated to coitus without affection because of its ease of occurring and the fewer demands it places on one's personality. But this danger can easily be overcome if we merely teach our youngsters that although both forms of coitus can be rewarding, coitus *with affection is most often a superior form of pleasure. In short, I think you are unfairly picturing me as proposing a "philosophy of blame and guilt". I am proposing a hierarchy of values and that is all.*

If you grant me the society in which people are brought up the way I think is preferable then I do not think there would be any problem of sex without affection driving out sex with affection. Such a problem is typical of our society because, as we agree, we stress the taboo on sex without affection and I think such might be possible in another society if it stressed the pleasure seeking activities and failed to build a socially accepted hierarchy of values in which sex with affection was preferred. I do not know if we really disagree on this last point. Your statements of what you would advise someone are very much the same as the advice I would give someone. I thus think that perhaps we've been fighting straw men—you fighting a position of 'blame and guilt' and I fighting one of 'no preferred type of coitus.' Whereas in reality neither of us held these positions. Perhaps we're both just anxious for a verbal battle and thus didn't fully see each other. I certainly do not favor a "blame and guilt" approach to coitus without affection—I often said it should be viewed like candy or quiz shows, a good but one that is minor. However, I still think that by telling people that A is better than B, you will make some people feel bad when they habitually practice B. But this is unavoidable for we must pass on our values, we must have our preferences and thus we must have some sort of social norms. Perhaps this is where we differ, perhaps you feel that we need not have any norms in the sexual sphere; but I don't think you do for your advice that you said you'd give shows a clear preference; and besides as I said in my last letter, we can have loose norms but to live together we must have some set of norms, some set of preferences, some hierarchy of values. Maybe you thought that when I said that guilt may result that I was favoring a "blame and guilt" approach—I wasn't, I was merely saying that to some extent it seems that guilt is impossible to avoid as long as we have social norms. With these norms, I am not afraid of coitus without affection driving out coitus with affection. That is only likely to occur when pleasure is stressed by our norms too much. Acceptance of coitus without affection as the lessor of two goods would not be overemphasis and would be perfectly acceptable to me. What I object to is the stressing of pleasure by making it "forbidden fruit" or by making people seek it as the ultimate good—both are to my mind erroneous positions. To place coitus without affection in a hierarchy of values which places coitus with affection much higher up, is my desire. I would pass these values down with the same conviction that I have that food is better nourishment than candy and Shakespeare better than Dragnet; but that at times it is good to eat candy and listen to Dragnet. Is this where we differ? Surely you would not refuse to take a position on such issues—surely you would say how you feel and what you believe to be right; then where do we really differ? Perhaps our difference is only that as a psychologist you're more desirous of stating things in terms of the individuals own decision right, and as a sociologist I like to emphasize the place of social norms in such decisions. Given a specific situation would we judge it differently? I can't think of a good test case, off hand, can you?

ALBERT ELLIS, Ph. D.
PARC VENDOME
333 WEST 56TH STREET
NEW YORK 19, N. Y.

March 23, 1957

Dear Dr. Reiss:

I quite agree with you that, by now, we have threshed out most of our differences and seem to be in better agreement than when we started. You, apparently, do not want to condemn all or most non-affectional coitus; and I do not want to say that all or most sex without love is better than affectional sex.

I will admit that I may be unfairly picturing you as proposing a philosophy of blame and guilt. I feel, however, that while you are not in this category, the great majority of other individuals who are concerned about sex without love definitely are.

We still have different points of emphasis on just how satisfactory unaffectional sex relations can be, and just how much stress should be placed on telling people how much better affectional sex acts are. But our differences in these respects do not seem to be too serious.

I think you are right when you say that I, as a psychologist, tend to look upon things differently than you, as a sociologist. I feel that social norms are most important, in that they exist, and have to be realistically acknowledged as existing. But I teach my patients, very often, to admit their existence—and then, as quietly and effectively as possible, personally to subvert them. Not, mind you, emotionally to rebel against them, or to become a martyr to such rebellion; but merely to fight their OWN tendencies to conform, and quietly to do pretty much as they damned please. You would probably like to see social norms changed, so that people would not individually have to fight against them. So would I; but I realistically have to face the fact that, at the moment, they are NOT being changed, and people are suffering from this. Therefore I teach these people how not to suffer.

I notice that you are down for a talk at the Eastern Sociological Society. which I shall try to get to. If you have any time available while you are in New York, you (and your wife, if any) are welcome to have dinner with my wife and I on Saturday or Sunday, April 13th or 14th. Just let us know a few days in advance if you can make it.

Sincerely,

Albert Ellis

Comments by Albert Ellis on His March 23, 1957 Letter

My rather extreme individualism came out in this letter to Ira. I accept social rules and regulations *in general* and hope that individuals acknowledge their existence. But I advocate that they subvert these social norms when they personally wish to do so—*and when* they can get away with their moderate rebellion and not be too punished for their rebelliousness. I held that the too rigorous might loosen up—for the benefit of all the rebels and also the nonrebels.

Today, I would take a more relaxed and "moderate" position. I would still say that the social rules, especially the sex standards, tend to be too rigorous and too rigid for many of the people much of the time. But I would also acknowledge their advantages—which are usually considerable. Therefore, I would advocate an investigative and experimental attitude by those who object to these rules and do some rebelling against them. Then these objectors and dissidents would *try out* rules that *seem*—yes, seem—better for them and perhaps for others and experientially *discover* their advantages and disadvantages. If they somehow monitored their findings and published them, *general* rules of conduct might be changed. Of course, such possible changes might be more liberal (as I would like it) or more conservative (as others might like it)!

COLLEGE OF WILLIAM AND MARY
WILLIAMSBURG, VIRGINIA

March 28, 1957

Dear Dr. Ellis:

We seem to be coming more and more into line with one another on our "symposium" topic. Perhaps we can iron out one more point. You mentioned that because norms change too slowly you teach patients to subvert them with tact and caution and suggest that I would prefer to wait until we had a new and better social norm. Again, I don't think our differences here are large. Norms, of course, do not change overnight, they change slowly and predominantly as a result of changes in social and cultural structure which encourages more and more people to violate the older norms. I go along with such behavior whole heartedly and think that this is one way that the older norms can be seriously weakened and new ones accepted. Sooner or later, several million people are going to realize that they have all been subverting the formal norms and living in a situation of pluralistic ignorance. When this happens, as it seems to be happening, due to Kinsey's publications, etc., many people join the newer norms and other people become more open in their violations of the older norms. As this process continues thru time, a new social norm becomes accepted. I think this kind of 'cautions violation' is fine but I have one qualification. <u>I would want the violator to have in mind a standard of behavior which he prefers to the one he is violating: I would not want him to violate the old standard just for impulsive behavior</u>. It is for this reason that I feel if there is a clear conception of the relative position of sex with and without affection it can be tremendously important—it can give a person a means of organizing his behavior and attitudes, instead of becoming anomic and a victim of his impulse. To allow that to happen would be to just substitute one tyrant for another, impulse for a disliked or conflict-ridden norm. I would prefer to substitute a personal standard to replace the formal social one, a personal standard that suits the individual. If this happens to differ from my own values, then I will be sorry but will accept it much quicker then mere impulse behavior. I would think that we don't differ too much here either for you did say you would tell your patients how you felt and ask them to examine your views to see if they fit themselves. This I would accept. Perhaps you did not mean to substitute impulse for the old norm but I got that impression from your letter, when you said that you teach them to "do as they damned please." Besides this difference (if it is real) I would fully go along with your position on subverting norms.

You are right about my being in New York in April. My wife, Harriet, will be with me and we both would very much like to get together with you and your wife. After all this correspondence we're both very interested in meeting "the other party" in person. Our schedule is going to be very tight time-wise but if it is convenient to you we could meet you about 1 P.M. on Sunday April 14th. We plan to start back to Virginia at about 3 or 4 P.M. so we would have a few hours together. Let me know if this is acceptable to you.

ALBERT ELLIS, PH. D.
PARC VENDOME
333 WEST 56TH STREET
NEW YORK 19, N. Y.

April 2, 1957

Dr. Ira L. Reiss
College of William & Mary
Williamsburg, Va.

Dear Dr. Reiss:

Yes, we do seem to be getting closer and closer in the expression of our views. I certainly would agree with you that people should not go about violating norms impulsively—but only after they think about them and decide that they are silly norms and could be replaced by better ones. When I say that I teach people to "do as they damned please" I mean, of course, after giving some serious <u>thought</u> to what they please. Impulsivity has its points, but easily gets one into difficulties with oneself and others. And, conversely, if one does not think, and meekly accepts the views one is given, that gets one into trouble, too—though largely with oneself and the loss of one's potentialities for full living.

More of this, no doubt, when we meet on April 14th. The main purpose of this letter is to say that I shall be glad to meet you and your wife, Harriet, at 1 P.M. on the 14th. My wife, Rhoda, will not be able to make it then, as she is stage managing a dance program that day and will have to be at rehearsal about noon. If you would like to see her, however, I expect to attend the concert at 3 P.M., and can arrange to take you and your wife along. It will last until about 4:30. Let me know if that is OK with the two of you, and I will get tickets for you for the concert.

Otherwise, I can meet the two of you at one and stay with you until about 2:30. Let me know which of these plans you prefer, and where you would like to meet me. You can make it at my place; at the Statler; or where you will.

Sincerely,

Albert Ellis

Comments by Albert Ellis on His April 2, 1957 Letter
In this letter to Ira, I tone down my liberalism somewhat and imply that it can be thoughtless and impulsive, hence disadvantageous. Today, I would say, "Right On!"

References
Ellis, Albert. 1954. *The American Sexual Tragedy*. New York: Twayne.
———. 1958. *Sex without Guilt*. New York: Lyle Stuart and Grove Press.
Reiss, Ira L. 1956. "The Double Standard in Premarital Sexual Intercourse: A Neglected Concept." *Social Forces* 34 (March): 224–230.
———. 1957. "The Treatment of Pre-marital Coitus in Marriage and the Family Texts." *Social Problems* 4 (April): 334–338.
———. 1960. *Premarital Sexual Standards in America*. Glencoe, Ill.: Free Press.

Enter Ethics, Religion, and Publications
Letters from April 5, 1957 to September 7, 1957

2

Introduction by Ira L. Reiss

AL AND I MET AT THE Eastern Sociological Society meetings in a rather unusual way. I was presenting a paper at the convention in which I asserted that premarital sexuality with affection was increasing in popularity and would replace abstinence as our formal sexual standard. After the paper there was a question/answer period. The first question came from a man who asked: "What about sex without affection? Why haven't you talked of that? Isn't that also going to replace abstinence?" Back in 1957, even my position about trends toward sex with affection was considered rather shocking and so I knew only a very rare person would pose a question like that. After all our letters, I felt I knew the source of this comment. I smiled and asked the questioner: "Are you Albert Ellis?" He replied: "Yes, I am." After our many hundreds of written words, those were the first face-to-face words we exchanged.

After the session, Harriet and I met Al closer up and we went to see his wife Rhoda in her dance rehearsal at the Henry Street Playhouse, an off-Broadway theatre. We only spent a few hours together but it changed our relationship from the formal one of Dr. Ellis and Dr. Reiss to Al and Ira. Now, about forty-five years later, Al and I still work in the area of sexuality and still occasionally exchange our ideas. There never was any dispute between us about our commitment to our professional work.

When we talked in New York, Al told us about the wedding announcement he and Rhoda had composed in 1956 when they married. It was indeed quite unique for its day. The document contains a jocular resume of each of them and displays their somewhat unconventional attitude toward their union. They married just three months after meeting each other. Shortly after the wedding, on that same day, they each went back to work! And instead of a honeymoon, they planned a trip to meet Alfred Kinsey later that summer! This is an indicator that the nonconformist approach that was present in Al's sexuality ideas was also present in many other areas of his life.

Harriet and I were less unusual in our 1955 wedding style. We were innovative only in having a Reform Jewish Rabbi perform the ceremony, but also having an Orthodox Rabbi participate. This combination was a compromise between our wishes and Harriet's grandmother's desires. Also, Harriet and I knew each other seven months before we married. So, even though we thought we were avant-garde, it does look like Al was a bit more of a rebel and more daring in his actions as well as his beliefs.

Both Al and I had book publication plans that were in high gear at this time in 1957. Al had been in the field for a number of years before we met and had already published a good deal. I was working on getting a publisher for my first book *Premarital Sexual Standards in America*, having just finished a first draft of the manuscript. Al was trying to do the same for his book *Rational Psychotherapy* (which was actually published as *Reason and Emotion in Psychotherapy* [1962]). As my April 21, 1957 letter to Al indicates, Macmillan had read my manuscript and told me they were "enthusiastic," but they were sending the manuscript out to other readers. We comforted each other's anxieties by noting that pioneers are always handicapped in publishing because of the unconventional perspective they take in their work. But our preference was to be recognized pioneers, rather than rejected ones.

Al and I shared many of the same evaluations of other writers in sexual science. For example, I thought Pitrim Sorokin's *The American Sex Revolution* (1956) was biased and almost totally lacking in relevant data—in short, not a good piece of scientific theory or research. Al, in his April 28, 1957 letter, before hearing my opinion, wrote that it was "a holy horror of biased and unclear thinking." We also found that we agreed concerning our high opinion of Clark Vincent's writings on marriage counseling. So I believe the two of us had a similar way of judging—we valued people who were careful and logical in their reasoning and were thoughtful and flexible in their use of relevant data. This shared perspective trumped any contrary emotional feelings we had and thereby promoted a broad scientific approach in our work and in our interactions.

On ethics and religion, Al was once more the more radical person. I had given up my Orthodox Jewish upbringing, but I still was a theist and a member of Reform Judaism. Al was an atheist and felt that ethics were "grunts of approval and groans of disapproval" (June 2, 1957 letter). He believed that "a thoroughly workable system of ethics [can] be built on reason alone." His ethics were based on "consistent self-interest." My thinking raised more issues by asking how we decide on what is in one's self-interest and what is the most "reasonable" ethical position? I felt we needed to make some assumptions about the value of other people and about the worth of taking a broader perspective that went beyond just oneself. I didn't think one could argue that it was just self-interest that led to accepting assumptions about the worth of human life. To me there were basic value beliefs that seemed to go beyond self-interest. Also, the concept of self can be seen as focused on just one's bodily pleasures, or on all of humankind, and so self-interest can include anything and everything. Therefore, I argued, there is no single definition of self-interest and so that too needs clarification.

Even though I was not able to persuade Al to change to my perspective, he did grant me (August 21, 1957 letter) that there needs to be more than reason to build a system of ethics. He proposed a basic value "that human life is good" and then deduced other values from that. He still had a lot more relativism than I did, for he granted that a system of ethics could include an ethic that saw human life as valueless. To me, such an ethic would permit you to eliminate others at your whim and I saw ethics as concerned with how to get along with other humans. But I agreed with his view that an ethical system should allow for change when it is clear that pursuing a particular value was leading to more harm than good. I also agreed with him that organized religion today often seemed to lack the ability to change its ideas about morality, especially sexual morality.

I saw more of value in religion and had chosen for myself a branch of Judaism, Reform, that allowed for a more flexible ethical system of thought. But Al would call that sort of religion not "real" religion because it was so liberal. So by definition, religion came out as the bad kid on the block, even if a few "exceptions" were excused from his negative label. I think part of Al's negative feelings about religion were due to his fundamental belief that "self-blame and other-directed-blame [lead to] . . . all other negative, happiness-destroying feelings" (August 21, 1957 letter) and that religion promoted just such self and other blame. I differed only in that I would not assign the promotion of blame so completely to religion. We create our religion from our basic societal values and so there must be in our societal values plenty of other sources that are promoting self and other blame.

It was in the June 2, 1957 letter that Al first mentioned to me about the new Society for the Scientific Study of Sex (SSSS) that he was seeking to found. In his August 21 letter, he asked me if I wanted to be a charter member of this new organization. I agreed and became one of the forty-one people that Al and five others persuaded to join. In my mind, SSSS was Al's idea and he was the key force behind it. He had tried earlier in the 1950s to found such an organization but Kinsey would not support it and it didn't happen. It is not clear why Kinsey felt this way, but shortly after Kinsey's death in August 1956, Al tried once more. Five other people were his key collaborators in this effort: at the start there was Hans Lehfeldt, Harry Benjamin, and Henry Guze, and in 1957 they were joined by Robert Sherwin and Hugo Beigel. This effort succeeded and SSSS today is the oldest and one of the leading sexual science organizations in the country.

COLLEGE OF WILLIAM AND MARY
WILLIAMSBURG, VIRGINIA

April 5, 1957
Friday Afternoon

Dear Dr. Ellis:

Your suggestion about the concert sunday afternoon sounds fine to us. We'd like to see the concert and more so would like to meet your wife. If it's convenient for you we'd like to meet you at the New Yorker. I guess by the main set of elevators would be as good a spot as any. From there we can decide what we want to go or where to go. Okay? So, if we do not hear from you to the contrary we'll be looking for you at 1 o'clock on Sunday, the 14th at the New Yorker Hotel. We're both looking forward to meeting you and your wife.

I have enclosed a copy of the paper I will be presenting at the convention.

Since we'll be talking in person quite soon, I'll close this letter now. I think this letter has set some sort of record for brevity in our correspondence. But we'll make up for it in N.Y. See you then.

ALBERT ELLIS, Ph. D.
PARC VENDOME
333 WEST 56th STREET
NEW YORK 19, N. Y.

April 8, 1957

Dear Dr. Reiss:

Glad you and your wife can make it Sunday! I shall meet you, as you say, at the main set of elevators at the New Yorker; we can then go to lunch; and thence to the dance concert.

I expect to hear you give your paper—which I have not read yet, but which looks interesting. In case, for some reason, I can't make it at the session, and you have difficulty recognizing me at the elevators, I am reasonably tall, thin, wear blonde glasses, and shall probably be wearing a black speckled topcoat.

See you Sunday!

Cordially,

Albert Ellis

ALBERT ELLIS, Ph. d.
PARC VENDOME
333 WEST 56th STREET
NEW YORK 19, N. Y.

April 16, 1957

Dear Harriet and Ira:

Here is the wedding announcement I promised to send. As you can see, it's just a little different from the usual one!

Rhoda and I very much enjoyed seeing you in New York, even though we didn't have much time together. We are looking forward to the next occasion.

Cordially,

Declaration of Interdependence
Constituting Our United Status

PREAMBLE

We, Rhoda Winter and Albert Ellis, in order to form a more perfect Union, establish justice, insure domestic—and we mean domestic—tranquility, provide for the common defense, promote the general (and particularly our own) welfare, and secure the blessings of liberty to ourselves and (fond hope!) our posterity, do ordain and establish this Declaration of Interdependence, constituting our United Status.

We are hereby announcing our intention to marry—yes, m-a-r-r-y—each other. So that all our friends may forever after hold their peace, and so that we may save our precious typewriter ribbon by avoiding writing them individually about the gory details, we are herewith giving them due notice of the Happy, if not exactly Blessed, Event and its antecedents and consequences.

Article I

I, Rhoda Winter, being cortically endowed, with due acknowledgment to Rose and Bob Winter, feel a close kinship with this Declaration. I was raised, with few conventions, in historic Philadelphia by delightful parents, who gave me my swaddling liberties in theatres, art museums, libraries, and the Academy of Music. By high school days, weaned on dance lessons, I had advanced to joining Thomas Cannon's ballet group, then performing with a local opera company. I thus saw the Academy of Music from the stage side and wore out several pair of ballet shoes to reach soloist rank. Meanwhile, the high school ballet and Katherine Dunham's extension group offered opportunities to study and perform.

In between darning tights, I managed to be mid-eastern editor of the American Friends Service Committee High School News Letter; pick up scholastic honors; be a leading light at Fellowship House; and, in the West Philadelphia High yearbook, be voted most likely. Then two hot summers studying dance in New York City on the Carnegie Hall circuit, plus two summers of dance counseloring at camps, convinced me—"Off with the ballet shoes, Pavlova, you're a barefoot type, 'modrin dancer' now!"

Realizing what every girl should know, I tearfully left for the Wilderness—to wit, cosmopolitan, international Madison, Wisconsin—where I enrolled at the University's famed Dance Division of the Women's Physical Education Department. (Having subsequently been a good staff member, I always give the Department its proper title.) The sifting and winnowing which followed could fill a book or two, especially the years of work and study experiencing the time, space, and foree factors of movement and personality under Marge H'Doubler, Louise Kloepper, and Shirley Genther.

After living at an independent Co-op, hashing and cashiering at the Student Union, working at Wisconsin General Hospital, presiding over the dance group of the University, performing and choreographing at least eight dances per annum, and writing a thesis on dance as a reflection of social change, it was nothing to get the B.S., skip out on graduation, and fly to Europe to study dance with Mary Wigman in Berlin and take the International Dance Course in Switzerland. On my return to Madison, I was granted a two-year graduate teaching assistantship in the Dance Division. The Summer of 1953 was spent as visiting lecturer st Wisconsin, teaching dance techniques and composition. This culminated in my acquiring an M.S. and writing a thesis on Aesthetic Experience and Dance Composition: the Choreography of Feeling.

Deserting Wisconsin, I next unpacked my tights at Ohio Wesleyan University, where I taught, performed, and did production work. To add excitement to my 72-hour work week, I also performed and taught at Columbus. The Summer of 1955 found me teaching percussion and dance at Madison again, as well as creating dances for television.

Venturing to the BIG CITY in the fall of 1955, I obtained teaching jobs at Performing Arts High School, the Hanya Holm Studio, Ann Reno Institute, and community centers in White Plains, Queens, and Staten Island. Concurrently, just to keep my foot in, I also conduct classes in dance therapy st Manhattan State Hospital, study with Allan Wayne and Alwin Nikolais, and practice cartwheels in preparation for my Assistantship this summer with Margret Dietz at the Connecticut College School of the Dance in New London. Oh, yes, in between all this kinesthetic activity, I am toying—and I mean toying —with the idea of getting married. It should be fun—moving that potential into a kinetically (and sometimes frenetically) creative relationship, even at the risk of a few unwashed leotards!

Article II

I, Albert (sometimes yclept Al) Ellis, being over 21 and of almost-sound mind, herewith affirm that I was born of woman (Hettie Ellis, then legally married to Henry O. Ellis) in Pittsburgh, Pennsylvania, but had the good sense to emigrate to New York at the age of four. I was upbrought on the sidewalks of the West Bronx, but eventually abandoned Poe Park and the New York Yankees for the wilds of Manhattan.

I received my Bachelor's degree in administration and comic verse from the College of the City of New York; and, ultimately, an M.A. and Ph.D. in clinical psychology from Columbia

University. My first self-supporting ventures were in the fields of editorial work and business management; but then I got sorely bitten by the bug of sexuo-amative research and, as a by-product, wound up doing marriage counseling and psychotherapy.

I have worked as Clinical Psychologist at the Mental Hygiene Clinic of the New Jersey State Hospital at Greystone Park; Instructor in Psychology at Rutgers University and New York University; Chief Psychologist of the New Jersey State Diagnostic Center at Menlo Park; and Chief Psychologist of the New Jersey Department of Institutions and Agencies. I am a Fellow or Member of seemingly numberless professional organizations, including the American Psychological Assn., the Amer. Orthopsychiatric Assn., the Amer. Assn. of Marriage Counselors, the Amer. Sociological Society, the Amer. Anthropological Assn., the Amer. Group Psychotherapy Assn., the Amer. Assn. for the Advancement of Science, the Society for Clinical and Experimental Hypnosis, and the Amer. Academy of Psychotherapists. I also am a Diplomate in Clinical Psychology of the American Board of Examiners in Professional Psychology.

At present, I am in the full time private practice of psychotherapy and marriage counseling in New York City; and, in between sessions as it were, I carry on quite a bit—some people think far too much—research and writing. During the past ten years I have published over 100 papers in psychological, psychoanalytic, psychiatric, and sociological journals. My main books and monographs include: AN INTRODUCTION TO THE PRINCIPLES OF SCIENTIFIC PSYCHOANALYSIS (1950), THE FOLKLORE OF SEX (1951), THE AMERICAN SEXUAL TRAGEDY (1954), and NEW APPROACHES TO PSYCHOTHERAPY TECHNIQUES (1955). I have also edited two volumes: SEX, SOCIETY AND THE INDIVIDUAL, with Dr. A. P. Pillay (1953) and SEX LIFE OF THE AMERICAN WOMAN AND THE KINSEY REPORT (1954). My latest book, THE PSYCHOLOGY OF SEX OFFENDERS, written in collaboration with Dr. Ralph Brancale and Ruth R. Doorbar, is to be brought out shortly by Charles C. Thomas, Publisher. I am presently working on a series of volumes on techniques of psychotherapy, and in particular on a book on THE THEORY AND PRACTICE OF RATIONAL PSYCHOTHERAPY.

How, in the midst of all this activity, I shall ever get time to see my wife is a good question. But, as I keep telling my clients, that will be Rhoda's problem.

BILL OF RIGHTS

We, Rhoda and Albert, having first met through the kind intercession of Dr. Warner A. Lowe on the 17th of February, 1956, and having adequately diagnosed and psychotherapized each other in the interim, hereby declare our intentions to betake us to New York's City Hall on the 26th day of May, 1956, and thereat duly to take the plunge into legal matrimony. After the simplest possible ceremony, and a pioneering ferryride to Staten Island, we shall then hie us homeward so that (a) we can go about our regular Saturday afternoon business (time and patients wait for no man-woman relationship) and (b) can attend the Henry Street Playhouse Dance Concert that evening.

Honeymoon? Lord, no! More appropriately, we are looking forward, this August, to visiting Dr. Alfred C. Kinsey (no cracks, please!) at Bloomington, Indiana; the American Psychological Association convention in Chicago; and Rhoda's old friends (if they'll still acknowledge her) at Madison, Wisconsin. After that we shall give serious thought to vacating our present temporary residences for larger professional living quarters. (Anyone hearing of a sizeable, suitable apartment in the Columbus Circle area of New York please phone or wire pronto.)

AMENDMENT

To give all of Rhoda's and Albert's friends a chance to denounce the groom for selfishly removing Rhoda from circulation and (albeit a bit belatedly) to kiss the bride, there will be a postnuptial cocktail party on Sunday, June 10th, 1956 at Apartment 10B, 333 West 56th Street, New York City, from four to seven o'clock. All those who can borrow or steal a ride to New York are most cordially invited.

Comments by Albert Ellis on His Attachment to His April 16, 1957 Letter

My wife Rhoda and I enjoyed writing this announcement and were happy that it was well received by Ira and many of our friends and relatives. Rhoda and I had been living "in sin" for a few months before our marriage but to please Rhoda's more conservative parents and relatives, we decided to be legally married.

As can be seen by some of my books mentioned in this letter, I was already an established sexologist, as well as a clinical psychologist and psychotherapist at this time.

Rhoda and I fairly conventionally signed this Declaration of Interdependence and agreed to have a monogamous marriage. This was quite different from my open marriage with Karyl, my first wife, and from that with several other women with whom I had had nonmonogamous relationships. But I was "pretty sure" that Rhoda and I would "last" and quite sure that if we did not, the end of the world would hardly come. So I happily took the plunge!

April 21, 1957

Dear Al and Rhoda,

It's just one week since we saw you in New York and we're just beginning to recuperate. We did not get home till 3 A.M. monday morning and it was a tough job meeting three classes a few hours afterwards. But despite the hecticness of the N.Y. trip we both enjoyed it a great deal. We enjoyed our meeting with you and Rhoda and hope that next time we can get together more leisurely. This summer we plan to come to N.Y. We'll keep you posted and if you and Rhoda by chance should want to visit the Jamestown Festival, we'd love to have you.

We received your most unique "Wedding Invitation" this week. I've never quite seen anything like it. There's little danger of it becoming customary—for even though the printers might like the idea, most people wouldn't have such backgrounds. By the way, did you and Rhoda actually go about your regular duties after the wedding on the day you were married.

How are things going with you and Rhoda—any new leads on the publication of your "Rational Therapy" book? Any new dance concerts planned for the Henry Street Playhouse? The only new event in our own existence here was our joining of the Reform Jewish Temple in Newport News. Up until this point we had been unaffiliated, but in the last few years we investigated Reform Judaism and found it remarkable free from the dogma's and incongruities of most organized religious sects. We've gone to services about a half dozen times in the last few months and enjoyed them also. This is the sort of liberal type of religion that you get in sects like the Unitarians and Congregationalists in Christianity. Before I go any further in this discussion I should ask you and Rhoda what your own religious connections are. I take it your background is Jewish but that leaves many possibilities open and in order to avoid offending anyone I ought to check with you more closely.

I still have my fingers crossed concerning my book on sexual standards—Macmillan wrote me the other day saying the reaction by their first reader was "enthusiastic" and that it was being read by others now. As a pro at this business maybe you can give me some advice—how long does it usually take from the time a manuscript is accepted to actual publication? I realize it varies but what has been your own experience? I would guess about 6 months—is that right? One other thing that I'm curious about—you said that Doubleday at the last minute bowed out and Boni published your "Folklore" book. Wasn't Doubleday bound by a contract? How did they get around that? As you can see, this is my first venture into publishing of this sort and I'm a little up in the air as to what it will be like.

I read a small but interesting new book this afternoon <u>Marriage: Past and Present</u>, it's a debate between Briffault and Malinowski that took place over the

BBC in 1931. Montague edits the 90 page book and tries to referee. It has historical interest mainly and is also worthwhile in that after Malinowski criticizes Briffault and Montague criticizes Malinowski, it is easy for the reader to criticize everyone for it is a most unscientific book, although a stimulating one. You might find it interesting.

I better close now—Harriet's birthday is this wednesday and I have to figure out something unusual to surprise her with. Write us when you get time and keep us informed of what is new

ALBERT ELLIS, Ph. d.
PARC VENDOME
333 WEST 56th STREET
NEW YORK 19, N. Y.

April 28, 1957

Dear Ira and Harriet:

 Rhoda and I are looking forward, if possible, to accepting your cordial invitation to come to the Jamestown Festival; but it is questionable that we shall be able to make it—particularly since Rhoda will be away the better part of the Summer teaching at the New London Connecticut College Dance Festival and I shall be commuting there, or she here, on weekends. We certainly hope to see you in New York, however—especially if you can make it in June, before Rhoda leaves for New London. Since we have two apartments, please feel free to use one of them if you want to stay overnight or for a few days.

 Yes, we actually followed the Wedding Announcement to the letter and went about our regular duties after the ceremony. Time and patients wait for no man or woman.

 We are exceptionally busy, as usual. Rhoda may appear in a concert at the Playhouse in May, and is still helping direct the childrens' productions. I have no word on my RATIONAL THERAPY mss. yet, but just signed a contract with Crown to publish my book, HOW TO LIVE WITH A NEUROTIC. I also may do a second piece for Esquire on sex and heart disease. They wanted me to do a very scary piece; but I told them I had no intention of making more patients—I have enough already. So it remains to be seen whether they will like the less scary piece they asked me to do.

 As you have estimated, it generally takes six months or more after acceptance of a manuscript for its actual publication. Occasionally, they rush it through in less time. Macmillan, of course, is an excellent publisher; and if they are enthused, so far, about your mss., I think there is a good chance they will ultimately take it. And I am sure that you will not have any difficulty with them. The reason I had difficulty with Doubleday was because they had an original contract with Charles Boni to publish the book; and, rather than fight them on this contract, Boni agreed to have it published under his name rather than theirs. He could have chosen to fight, no doubt, but decided not to do so.

 I read MARRIAGE: PAST AND PRESENT, and reviewed it for Marriage And Family Living. I thought it quite interesting and felt that although Briffault's heart was in the right place, Malinowski got the better of the debate. I also reviewed Sorokin's THE AMERICAN SEX REVOLUTION at the same time, and found it a holy horror of biased and unclear thinking.

I just received a copy of Clark Vincent's READINGS IN MARRIAGE COUNSELING; and, quite apart from the fact that he uses some of my material, I thought he did an unusually good job of compilation. He includes some liberal views that most of the other books of this nature omit.

Rhoda and I are nominally Jewish; and I attended, in fact, a Reformed congregation in the Bronx when I was a child. We both became atheists, however, at a tender age and have kept away from entangling alliances ever since. My impression, like yours, is that many Reformed Jewish congregations are no more deistic than the Unitarians, Humanists, or Ethical Culturists; and I note that many of the liberal Protestant clergy are rapidly giving up god in favor of ethics. A fine move, I believe; but I still like to call ethics ethics and religion religion.

Rhoda joins me in sending our best wishes to Harriet on her birthday. And the best of luck to you on your manuscript at Macmillan's!

Cordially,

Comments by Albert Ellis on His April 28, 1957 Letter

Doubleday didn't want to publish my book *The Folklore of Sex* under their own name because they were afraid that several of their own authors, whose material I quoted in the book to show how sexy or prudish it was, would take offense to my criticism. I doubt very much whether they would have objected, since it publicized their writing. But though Doubleday printed and published the book, they put it under Charles Boni's name. Then they didn't push it very hard and it sold very few copies. Ten years later, in 1961, I arranged for a revised edition to be published by Lyle Stuart and a paperback edition to be published by Grove Press—and it then sold very well for the next several years!

In this letter to Ira, I was somewhat dogmatic and punitive, which theoretically, I oppose. I *completely* throw out religion and do not see it as having *any* place in ethics. Actually, it has some ethical correlates—and, maybe, even a few "good" ones. I still agree that ethics are largely based on enlightened self-interest and that theoretically suffices. But humans do *not* always think intelligently and they often *are* disturbed. Therefore, they require, for ethical behavior, what REBT calls unconditional self-acceptance *and* unconditional other acceptance. In fact, one could hypothesize that both these philosophies and behaviors have overtones of a value-oriented Belief-Emoting-Behaving System that is basically and intrinsically ethical-religious. It is motivated by the urge for human survival *and* self-actualization. As Kant noted, it may not be empirically proven or falsified, but it had damned well better be *chosen!*

ALBERT ELLIS, PH. D.
PARC VENDOME
333 WEST 56TH STREET
NEW YORK 19, N. Y.

June 2, 1957

Dear Ira and Harriet:

Rhoda and I are sorry to hear that you probably won't get to New York this Summer. We do have two apartments—one being my office, which I used to live in before marriage, but which now serves as an office, and the other one our regular apartment. So if you ever get around here for a few days, you're welcome to use the office. I use it myself between 10 AM and 11 PM; but, aside from that, it's all yours!

I'm not sure what the publication date is, as yet, for HOW TO LIVE WITH A NEUROTIC; but I imagine that it will be next Spring. In the meantime, I have a book of my INDEPENDENT columns coming out this Fall. And I'm now finishing up the manuscript of a sex manual to end all sex manuals—tentatively titled EVER THE TWAIN SHALL MEET. So you can see that I keep busy. The Esquire piece is scheduled for publication around October.

I have finally polished off most of our correspondence on sex and love in three INDEPENDENT articles, and am sending you the issues for your files. I'm just about sick of the damned subject by now!

I still think that Macmillan's keeping your mss. so long looks quite good. By all means let me know if you hear anything definite from them.

About God and Ethics: I cannot see why ethics need any justification other than man, his healthy survival, and his happiness. Why does any set of rules have to have any set of meaning outside one's own grunts of approval and groans of disapproval? All rules, it seems to me, are only sensible in so far as they make for better living; and I cannot imagine what other purpose they should have. I also cannot imagine how any set of rules could possibly have any ultimate meaning beyond the purposes of those who make and live by them.

I believe that Immanuel Kant, who brilliantly proved in his CRITIQUE OF PURE REASON, that one could never prove the existence of any deity, turned coward and unwarranted pessimist when he claimed, in his CRITIQUE OF PRACTICAL REASON, that one needs a God as a necessary assumption for a system of ethics. If, indeed, man cannot build a system of workable ethics entirely on the basis of reason, intelligence, and common sense, then I would say that he does not deserve to survive, and that he and his hypothetical Gods might just as well perish.

It is my firm belief, however, that not only can a thoroughly workable system of ethics be built on reason alone, with no supernatural assumptions of any sort; but that such a workable system can ONLY be built on a thoroughly objective, scientific, and reasonable basis. And that basis, at bottom, is nothing but consistent self-interest. Starting from self-interest alone, and following this through to its logical conclusion, I feel that no intelligent and undisturbed human being would commit unethical acts—by which I mean acts that needlessly harm other human beings or take advantage of minors—because if he does so, he would just be sabotaging himself and his loved ones—which would be against his own self interest.

We once started to discuss this, on the way to the Henry Street Playhouse, but never finished our discussion. Sometime when we have a few hours to spare, I hope to discuss it with you and Harriet. In the meantime, I am quite willing to go along with various reformed and liberal conceptions of God as being less idiotic and disturbed than the usual orthodox conceptions. At bottom, however, I still feel that they are based on unnecessary assumptions and hence are superfluous. The only God that makes any sense at all to me is Spinoza's pantheistic God, which is equivalent with Nature or Reality. But since I am certain that we have Nature and Reality, what the devil do I have to make additional assumptions by calling them God?

Life has largely been work and a series of professional meetings here. Most important have been the founding of a new society, the Society for the Scientific Study of Sex, on which I shall send you literature soon; an all-day conference which I helped arrange for the American Association of Marriage Counselors on June 7th, in honor of Kinsey's memory; and a workshop this Summer and an annual meeting this Fall, both of which I am helping with for the American Academy of Psychotherapists. So life seems to be toujours meetings, whathehell, whathehell!

Rhoda, who will leave for Connecticut July 5th, sends her best, as do I.

ALBERT ELLIS, Ph. D.
PARC VENDOME
333 WEST 56TH STREET
NEW YORK 19, N. Y.

August 21, 1957

Dear Ira and Harriet:

I can well see that both of you have been too busy for corresponding. My own position is disgustingly similar: since I have been to Madison, Wisconsin for a workshop on psychotherapy, and then to the Dance Festival at New London (where Rhoda has been for the past six weeks). Now I am preparing to give two papers at the APA meetings in New York which begin next week. Rhoda, meanwhile, has put me to shame by dancing, teaching, lecturing, being impressario, etc., etc., and will probably have to take a few weeks to make up for the overwork she has just gone through.

I can well appreciate your publishing difficulties, particularly because they sound somewhat like mine on my rational psychotherapy manuscript. Like you, I find that publishers are reluctant to touch it because it differs from existing work in the field—which apparently means that pioneers are notably handicapped in publishing anything. But, again like you, I am reworking the material myself, in the hope that eventually I will find someone who is interested in putting it out.

There are no reprints of the Wylie article available, to my knowledge; but I am sending you, under separate cover, a battered copy of the book in which it appeared, this being the only extra copy I have around.

Pillay died over a year ago; and the Int. J. of Sexology consequently no longer exists. His coworkers are still publishing the Journal of Family Welfare, its companion journal; but this is not up to the old one.

The Society for the Scientific Study of Sex is now well launched, as the enclosed circular will show; and <u>if you would like to become a charter member</u>, just drop me a note, and that will be all the formality required. It is going full blast, and we are getting some very good people to accept charter membership. No attempt has been made as yet to get general members.

You are quite right in saying that I was wrong in stating that man can build a system of ethics entirely on the basis of reason. Behind every ethical system, there has to be at least one basic value—such as the value that human life is good—and from there on one can rationally deduct various sub-values. What I really should have said is that we need no extra-human assumptions or values from which to originate our moral codes. All religions worthy of the name, in my estimation, make extra-human assumptions; and assumptions, moreover, which are in principle non-veridical. Moral codes, however, are verifiable on a contingency level. That is, we can say that *if* this is the kind of a world you want to live in, then you have to act in such and such a manner. If, however, your value system is quite different (if, for example, you believe that human life is valueless, and that people should not try to live together peacefully and non-anxiously), your moral code will also, if it is rational and logical, tend to be different.

An ethicalist differs from a religionist, moreover, in that while he has faith in one sense of the word (in the sense that he has some values), he has no absolute or arbitrary or immutable faith. He believes, fundamentally, in a given system of values because he tentatively thinks or surmises or guesses that it will work better than another system. But if factual evidence arises to make him think, surmise, or guess differently, he has little difficulty in changing his basic value system The religionist, on the other hand, believes irrevocably and finally in his value system and can never change it except to give up religion and become a non-religionist. Thus, my own basic ethical value is that of enlightened self interest; but if you convince me, as you haven't done as yet but as you may well do one of these days, that a workable system of ethics cannot be derived from this main principle, I shall either change the principle or add others to it. If you convince a religionist, however, that there are no extra-human entities, he cannot merely change to another religion, but must become (whatever he chooses to call himself) a non-religionist.

Confusion arises because of the various meanings which are normally given to the word "religion". Thus, Webster's Collegiate Dictionary defines it as follows: "1. The service and adoration of God or a god as expressed in forms of worship. 2. One of the systems of faith and worship. 3. The profession or practice of religious beliefs; religious observances collectively; pl. rites. 4. Devotion or fidelity; conscientiousness. 5. An awareness or conviction of the existence of a supreme being arousing reverence, love, gratitude, the will to obey and serve, and the like." All these definitions, it seems to be, are accurate, except possibly the fourth: since, under its usage, a conscientious atheist would be a religionist. Personally, though, I would even tend to accept the fourth definition, and call anyone who is over-conscientious, devoted, worshipful, obsessed, or possessed a religionist. And anyone who is consistently objective, dispassionate, open-minded, scientific, and rational, I would tend to call a non-religionist. Many atheists (such as the Marxists) I would therefore consider actually to be religionists; and many "believers," such as a good many Humanists, Ethical Culturists, Unitarians, etc., I would consider as non-religionists.

On your point about people who get away with crime: Doubtlessly, a few such individuals exist. Actually, however, damned few: since even those, like Lucky

Luciano, who seem to pull this trick are almost certainly heir to all kinds of anxieties, fears, guilts, hostilities, etc., which are, to my mind, simply not worth the gains made by a criminal life. The only truly happy human beings, by my definition, are those who have very fleeting and absolutely minimal anxiety and hostility, or self-blame and other-directed-blame (from which two emotions I can derive all other negative, happiness-destroying feelings). It may be theoretically possible to imagine a society in which, while getting away with antisocial activities, humans would still be non-anxious and non-blaming; but I doubt it. When I say, therefore, that people inevitably hurt themselves by engaging in criminal activities, I do not only mean that they risk exposure, jail sentences, ostracism, etc.; but, more importantly, they also inescapably bring on their _own_ anxieties and hostilities.

One other point: assuming, for the sake of discussion, that we could imagine a society in which a person could be unethical and still not harm himself in any major way, it should be obvious that, by virtual definition, such a person would have to be a rare exception in his community. For if _many_ such people existed, almost everyone would soon be harming everyone else, and chaos (as Immanuel Kant pointed out many years ago) would result. Ergo: at the very most, while granting the possible validity of your objection to a system of ethics based on enlightened self-interest, I would have to change my formulation to read: Enlightened self-interest is the best ethical system for _most_ humans to follow if they reside in communities in any way resembling those societies which have thus far been historically extant.

Your point, however, is very well taken in a practical sense—and that is that when one resides in a culture or sub-culture (such as the business sub-culture of our own society) where the _general rule_ is for people to be more or less immoral, or at least very loosely moral, _then_ one had better not try to be Jesus Christ, or else one will get more hurt than if one were just about as immoral as the rest of one's fellows. Thus again: if one is fighting in a war, one had better not adopt the moral attitudes toward one's enemies that one would adopt toward one's friends and associates in peacetime. Certainly, there are these practical exceptions to moral rules. But I still think that the general rule, for most people in most cultures, is a good one.

Rhoda sends her best, and joins with me in hoping that you will let us know when you are next around these parts so that we can spend some time together.

Cordially,

Comments by Albert Ellis on His August 21, 1957 Letter

I again go too far, in this letter to Ira, in calling *all* humanists, ethical culturists, unitarians, and so on "nonreligionists." Most of them are not absolutist or dogmatic religionists but they still may have profound faith in some aspects of what we usually call "religion."

I am happy to see that I incorporated self-blame and other-directed blame, or lack of human damnation, in this early REBT philosophy of maximum mental health and minimum self-sabotaging. Rigid self and other damnation may not be the only sources of people's disturbances, but REBT still sees them as prime sources.

I would say today that enlightened self-interest and enlightened other-interest are probably the best ethical system for most humans to follow. But let's test them out and see!

ALBERT ELLIS, PH. D.
PARC VENDOME
333 WEST 56TH STREET
NEW YORK 19, N. Y.

Sept. 7, 1957

Dear Ira and Harriet:

The APA meeting has come and gone; and, as usual, it was a seven-ring circus. Loads of papers, many of them interesting, many humdrum. I had one on neurotic interaction between marriage partners, in a symposium on marriage counseling; and one on outcome of employing three techniques of psychotherapy. Both actually beat the drum for my new techniques of rational psychotherapy; and both raised hearty discussions, much of which (to my surprise) was favorable. I shall send along reprints ultimately; so you can judge the material for yourself.

Yes, Pillay was the financier, as well of the editor, of IJS; and without his backing the journal is a dead duck.

If you want to speak of liberal and orthodox religionists, that's all right by me; but I can't see any essential difference between most of the liberal boys and the non-religious ethicalists. Or, if there is a difference, it is this difference to which I object. The only real objections I have to the liberal religionists—assuming that they are not essentially different from ethicalists—are (a) they still hold on to much of the traditionalism of the orthodox boys, and hence become confused with them; (b) they are taking a perfectly good word, "ethics," and redefining it so that it now means "liberal religion." Why use two words, if there is no essential difference between them, when the first of these words, ethics, is quite clear and can hardly be confused with orthodox religion?

By the same token, I object to Freud's using the word "sex" which essentially refers to psychophysical stimulation leading to genital arousal and satisfaction—to cover (a) sensuality (which is largely non-genital) and (b) love (which is largely non-sexual). If we have three fairly discrete words here, why not use them properly, instead of confusedly Intertwining them?

I agree that Lucky Luciano may not be anxious; but I cannot see his not being hostile. Anyone who needlessly keeps doing in others, and does not get anxious about some kind of retaliation, will almost inevitably have to hate those he hurts; and such hostility, I feel, is self-damaging, and not worth the monetary or other gains involved.

You are perfectly right in saying that in a non-integrated society, one can break moral rules and often gain more than those who abide by them—at least in a financial or property sense. This is tantamount to saying that if one is perfectly logical in an illogical culture, one will doubtlessly suffer because of one's culture. Agreed. But how do we get to a more logical culture unless individuals make at least some small sacrifices in getting there?

Since society, as you well point out, has not as yet shown any great capacity for sanely solving many moral questions, one often has to be "immoral," from the definition of one's society, in order to be happy. So much the worse, then, for society. I definitely teach my patient's to scout many of society's laws, particularly its sexual codes. But if they are inconsistent with their OWN system of values, they will definitely be confused and unhappy—even though they get away with a hell of a lot, as far as society is concerned. Lucky Luciano, I would say, is one of the many who is inconsistent with his own values. One the one hand, he probably believes that he is doing the right thing by directing thievery, murder, etc. But, on the other hand, he probably wants to have self-respect, kindly feelings toward others, etc. Unfortunately, he cannot <u>really</u> have both; though, by repressing or dissociating some of his own behavior, he may neurotically believe, on one level, that he is a great guy while, on another level, almost certainly think himself a louse and/or (as I said before) hate others. The fact that his anxiety and/or hostility is not necessarily conscious does not prove (at least to me) that it does not exist.

In other words: I quite agree with you that, as far as society's laws are concerned, it is fairly easy for any intelligent individual, and even a Lucky Luciano, to flout these laws and to gain greater worldly goods and other advantages than many individuals who rigidly follow such rules. But I do not see that any human being in almost any kind of a society can blithely go about needlessly hurting others and <u>really</u> not suffer serious personality damage thereby.

There should certainly be no exorbitant fees or other strings attached to the SSSS; so I am putting you through as a charter member. You will hear a little later on, through official channels, whom all the other charter members are, and what the society is going to do.

Cordially,

Comments by Albert Ellis on His September 7, 1957 Letter

In this letter, I still tend to put all religionists in the same category—but this is rigid and inaccurate. I now differentiate between religionists who believe in damning and punishing gods—whom I would largely deem unethical and emotionally disturbed—and those who believe in undamning forgiving gods—who I would deem less disturbed. But even then, there could be exceptions to my generalizations!

Today, I would still go along with my client's who are sexually unconventional—as long as they are willing to take the risks and penalties of being so. And I still would say that the Lucky Lucianos of the world may be remarkably little guilty and self-damning but would almost always be hostile and other-damning.

References

Ellis, Albert. 1951. *The Folklore of Sex*. New York: Charles Boni.
———. 1962. *Reason and Emotion in Psychotherapy*. Secaucus, N.J.: Citadel.
Reiss, Ira L. 1960. *Premarital Sexual Standards in America*. Glencoe, Ill.: Free Press.
Sorokin, Pitrim. 1956. *The American Sex Revolution*. Boston: Sargent.

Therapeutic Ideas and the Launching of a Sexual Science Organization 3
Letters from November 10, 1957
to November 3, 1959

Introduction by Ira L. Reiss

TOWARD THE END OF 1957, our dialogue on ethics moved into the mental health area. It became clear that once again the difference between us was that Al focused more on the individual and his or her goals and preferences while I focused on how such individual goals were strongly structured by the family, religious, political, economic, and educational forces in the society. Because of my societal interests, I gave top priority to working to changing these social grouping. To me, society shaped our reflexive self and a thoughtful plan for self-change should acknowledge that. But Al saw "healthy" individuals as setting their own goals and being able to resist the pressures of others. In fact, he went so far as to say in his November 10, 1957, letter, that "[w]hen humans need some generalized socially agreed-upon 'purpose' to give them 'peace of mind' they are then, in my estimation, emotionally disturbed." Al was stressing the rational and the scientific ability of the individual to shape him- or herself.

In November 1957, Al sent me a copy of his new book *How to Live with a Neurotic*. As part of our discussion of this book, he did grant that we live in groups and that our self-development is tied to others, but then he added his qualifier: "My anarchistic bias, however, is to change the world so that we become much less a part of others and much more individualistic." Compared to Al, I sounded like a communitarian. But I also argued, though more moderately, against conformity to customs. I stressed careful examination of customs to see if they fit with our individual views and values. So I, too, was quite willing to pay the price of a more individuated society rather than live in a more conformist world. But I wanted to recognize more of the immense pressures from others that exist and that inevitably exert an emotional as well as a rational force. Those pressures are what we cannot eliminate but can reshape. Most other people in America in 1957 were to the right of me. But Al was one of the few who was on my left.

At their eighteen-month wedding anniversary, Al and Rhoda give an update on their relationship. This document is included in this chapter of letters. Rhoda and Al both led extremely busy lives as this marital update indicates. Rhoda was into dance, although some of her work involved dance therapy at Manhattan State Hospital and thus overlapped with Al's focus. Al's publications continued at a fantastic pace. How he could see his usual sixty patients a week and write all the books and articles he did has always amazed me. I wondered when Al and Rhoda had time to talk.

By July 1958, he was busy on his *Encyclopedia of Sexual Behavior* and asked me if I wanted to do an article in that book. I accepted but my energies were still focused on securing a publisher for my book *Premarital Sexual Standards in America*. I finally achieved that goal in May 1959. After a number of rejections by other publishers, the Free Press of Glencoe, Illinois—one of the finest academic presses in the country—did enthusiastically accept my book. I was thrilled. But keep perspective on this—bear in mind that Al had put out at least three books during the few years that I worked on this one. I also had received a job offer from Bard College in Annandale-on-Hudson, New York, and had decided to leave William and Mary by the fall of 1959. So my life was in motion, too, but at a lower-frames-per-second-pace than Al's was.

In November 1958, a very important event was to take place in New York City. The Society for the Scientific Study of Sex (SSSS), founded in 1957, held its first convention. Al and I planned together with our wives to spend some time with each other during that 1958 meeting. But events intervened. In early September 1958, Harriet discovered that we were pregnant and decided not to go to the meeting. In addition, as it turned out, shortly before the November meeting, Al and Rhoda got a divorce. The breakup was friendly and they were both so busy that it seemed to be relatively easily absorbed. His explanation was: "We have rationally decided that our common interests and goals in marriage aren't extensive or intensive enough to warrant our going ahead with the marital relationship" (November 2, 1958 letter).

The first SSSS meeting was a good one with a focus on abortion issues and discussions of "frigidity." Hans Lehfeldt and Al, two of the founders of the organization, were among those who gave papers. Most of the other papers were given by medical doctors. Some of the presenters were very well known, for example, Clyde Kiser, Christopher Tietze, and Alan Guttmacher, among others. The papers from this first meeting were published in the *Quarterly Review of Surgery, Obstetrics, and Gynecology* (1959). An announcement of the new SSSS organization was contained in this publication together with the names of the forty-seven charter members.

While at the meeting, I raised the issue with Al of the relation of science and value judgments. We surely had been batting this issue around in our letters. And the potential clash of science and values could be seen in the papers at this first meeting. For example, how one evaluated abortion and frigidity clearly reflected more than "pure" medical facts. Al liked this topic a great deal and he asked me to organize the December 1959 meeting of SSSS and to be the main speaker. That meeting of SSSS, like most of these early meetings, was held at the Barbizon Plaza Hotel in New York City. I was able to convince John Money and Walter Stokes to participate and the program was a success. SSSS was off to a good start.

At about this same time, after hearing that Al and Rhoda had split, I informed my sister who had been divorced just two years earlier. She and Al got together and went out for a brief period of time. But nothing more lasting came of it. I thought it would be great to have Al as a brother-in-law but it was sufficient to have him as an intellectual friend. Al had enough of marriage at that time. Also, winter was setting in and he disliked the cold weather and he was doing less in terms of social life. But he and I kept up our correspondence on the article I was writing for his *Encyclopedia*, which eventually came out in 1961. In addition, I was writing an article about love, where I proposed my "wheel theory of the development of love," which was published in May 1960 in the journal of *Marriage and the Family Living*. A good deal of the ideas and evidence on this "wheel theory of love" was in my first book when it was finally published in August 1960.

In the fall of 1959, I went off to Bard College—just one hundred miles up the Hudson River north of New York City. Our first child David had been born June 9, 1959, only eight weeks before we made this move. Al encouraged me to take this new position and stressed the advantages of being near to New York. Having been raised in the East, both Harriet and I did feel more comfortable in New York than in Virginia. I had applied to the National Institute of Mental Health (NIMH) for a research grant to study whether there is indeed a new premarital sexual standard evolving in America as I had described in my 1960 book. As I wrote Al (March 7, 1959 letter), I had started a research project using college and high school students to test my ideas and my scales for measuring changes in premarital sexual standards. This research would lead to several national research grants and be the basis for my 1967 book. It was Stokes, one of the SSSS charter members, who told me that the National Institute of Mental Health (NIMH) was willing to support sex research and that I should apply. But the interest at NIMH in sex research was more covertly than overtly supported. After I applied, I was advised by NIMH not to use the word "sex" in my title of the grant application—even though the research study was focused on sex. I went along with that and in the late spring of 1960 I learned that my grant had been approved. So sex and the government did work together even back in 1960, but only when done discretely. If only President Bill Clinton had learned this lesson of the value of discretion when dealing with sexuality!

Throughout this period, Al was as usual extremely busy but he told me that the fast pace didn't bother him. As he said in his May 10, 1959 letter: "Never an unfrantic moment!—but I really don't take things that seriously, so manage to get through without ulcers." I'm afraid I never learned how to be quite that calm, cool, and collected. I was a milder form of a workaholic, but I admired Al's ability to operate at such a seemingly frantic pace and enjoy it to boot.

ALBERT ELLIS, Ph. D.
PARC VENDOME
333 WEST 56TH STREET
NEW YORK 19, N. Y.

Nov. 10, 1957

Dear Ira and Harriet:

You two sound almost half as busy as we are! Rhoda is doing her dance therapy at Manhattan State Hospital; rehearsing for a production that is scheduled to go on early in December; teaching extra classes at the Hanya Holm Studio; working with two special people in the field; etc., etc. I am seeing my usual 60 patients a week; doing much reading preparatory to revising my text on rational psychotherapy; knocking off occasional articles and chapters for other works here and there; and so on. In between, we have a hectic week-end social life, which always keeps one step ahead of us.

You are right that many—though, I think, by no means all—liberal religionists include a belief in some kind of ultimate meaning and purpose. It is here, exactly, where I think they become unscientific; as, in my estimation, the only meaning and purpose of life is living itself—acting, doing, being. This meaning, I feel, is born with life itself and dies with life. Living things can GIVE their life all kinds of meanings; but these are personal, preferential, and non-generalizable.

When humans need some generalized, socially agreed-upon "purpose" to give them "peace of mind" they are then, in my estimation, emotionally disturbed, as they then need an arbitrary definition of someone else's "purpose" to give meaning to their own lives. When they need NO other meaning than the individual, preferential meanings which THEY THEMSELVES give, then they are truly invulnerable to arbitrary "meanings," since they have their own. This "meaning" may be, as the case varies, estheticness, scientific pursuit, athleticism, etc. But the "ultimate" meaning is still their own existence, their own life, and not some general goal or end. And their own ethics, as I have previously indicated, is in terms of their trying to live comfortably with others in order to achieve their individualized preferences, or at least not interfere with them.

The SSSS is coming along fine, and you will be receiving, one of these days, a complete list of Charter Members. We are first, however, going after some outstanding ones who have not joined yet.

You will be getting reprints of some of my new papers at the end of the year, when I collect them and send them out. The different forms of mental illnesses are describable within various frames of reference, and you are probably just as well off with your own as with anyone else's. I usually prefer to stick to descriptions of the characteristics of a disturbed individual rather than to different kinds of disturbance. My new book, HOW TO LIVE WITH A NEUROTIC, gives simple descriptions of some of the main characteristics of neurotics; and these, when taken to extremes, could just as well, in my opinion, characterize psychotics. I am sending you a copy under separate cover, and would be glad to have your reactions, especially the negative ones. The publisher rushed the book through at the last minute, so apparently there are a good many typographical errors as well; and any interesting ones that you find, by all means send notes on them along too.

Cordially,

Comments by Albert Ellis on His November 10, 1957 Letter

Again, I am a little too dogmatic about religionists. Perhaps not *all* of them have a belief in some kind of ultimate meaning and purpose, and some atheists, like the Bolsheviks, do seem to have a belief in ultimate meaning and purpose. It is people's rigidly holding such a belief and dogmatically—and sometimes punitively—trying to force other's to have it, too—this is what I object to.

When I say in this letter that people need no other meaning than the preferential meanings that *they themselves* give, I do not mean that they *cannot* choose more social or general meanings. Of course, they can. They can choose any meanings they wish to choose; and they also partly learn, in their social groups, to be prejudiced in favor of other-directed meanings *as well as* individually selected ones. But favoring ultimate meanings that they like, instead of being *too* beholden to socially imbibed meanings, still seems (to me) preferable. However, their largely selecting socially accepted ultimate meanings might *possibly* better preserve their human survival *and* their individual happiness. I guess there is no clear-cut answer here.

ALBERT ELLIS, PH. D.
PARC VENDOME
333 WEST 56TH STREET
NEW YORK 19, N. Y.

Nov. 24, 1957

Dear Ira and Harriet:

I certainly want to thank you for the trouble you have taken to go through HOW TO LIVE WITH A NEUROTIC, and for the quite pertinent comments that you make on it. Most of your main points I accept, though with my usual reservations. To consider them seriatim:

1. I agree that we have to be SOMEWHAT dependent on the attitudes and approval of others, at least in any society anything like our own; and I always make it clear to my patients that, at the very least, they must seek what I call "practical love"—meaning, the love of those whose aid and succorance they need (e.g., that of their parents when they are young, their spouses, their bosses, etc.). I go into this in detail in my manuscript on rational psychotherapy, but did not try to put in the ifs, ands, and buts in HOW TO LIVE WITH A NEUROTIC, partly because I did not stress the point of not needing other people's love too much in the book.

a. I certainly accept Mead's statements as descriptive of the way things are in virtually all human societies; but not necessarily as descriptive of the way they should be. I would call the delinquent boy who was guilty about cunnilinctus neurotically supersensitive because (i) he took his friend's opinions TOO seriously and (ii) his friends, in my estimation, unthinkingly accepted THEIR opinions from others. If my views on rational behavior become institutionalized in practice, Mead's picture of the world will radically change. I have, however, no idea that my views will be accepted to any large extent in the near future. I would agree with you (and Mead) that, because we are humans and live in societies, we are always via our self development part of others and thus there is no sharp line between ourselves and others. My anarchistic bias, however, is to change the world so that we become much LESS a part of others and much MORE individualistic. The pages in my book, incidentally, which show a greater acceptance of love were written two years ago, and there was not time to revise them drastically before the publisher rushed the book through the presses. Today, I would write them differently, and the book therefore would not be as contradictory as it is in parts.

b. I agree with you about the role-conflict issue, but stick to psychological rather than sociological terminology largely because I am a psychologist and the readers will be expecting this.

2. *Your point is entirely well taken. In my manuscript on rational therapy, again, I am spending a good deal of space making explicit the fact that reason or logic cannot provide truth, although they are of great help in sifting truth from falsehood. I also expect to make clear that the value system behind rational therapy—and it is, of course, a value system—starts with the <u>assumptions</u> that intense, sustained, and repetitive anxiety, hostility, guilt, depression, etc. are undesirable; and, given this assumption, it presents a philosophy and a technique for eliminating this kind and degree of negative feeling states. There will be no attempt to prove this assumption since, as you accurately point out, it must be either accepted or rejected on what is usually called "faith" but what I prefer to call "personal preference". This personal preference, in turn, is partly, but I believe by no means entirely, a result of the social conditions under which one is raised; but human beings, fortunately, can see through and rise above the social conditions of their raising—partly by use of reason and logic.*

3. As just noted, I quite agree that complete determinism, a la Hume, is an untenable position, and I certainly do not take it. I certainly feel that Hitler and Dillinger did not HAVE to be what they were; but I would say that—this time using YOUR more social concepts—their social roles were LARGELY determined by the milieus in which they were raised, and that therefore they would have enormous DIFFICULTY in not being what they actually were. Since the cards are so greatly stacked against them from the start, I hardly feel that it would be just, in the usual sense of the term, to BLAME them for finding things so difficult that they took the easier (for them) way out and became criminals. But this does not gainsay the fact that I heartily agree with you that man is not just a robot. If he were, then he would be determined to be neurotic WHATEVER he did; and rational therapy would be entirely useless.

I doubt whether I have any REVERENCE for science and logic; but I do admit that I have a hell of a lot of respect for man's—or at least some men's—ability to think and to rise above much (though probably never all) the nonsense which they are originally taught. I am inclined to agree with you that individualism, at least in today's world or that of the near future, is a relatively different combination of SOCIAL factors, since it would be folly if everyone tried to shed ALL his social learning to be individualistic. At the same time, it is difficult for me to see how one can share ALL his beliefs with others and still be anything of an individual. He would then be a PERSON, but hardly, in my sense, an individual. Of course, if he shared ALL his beliefs with SOME others, there are doubtlessly enough different beliefs to go round so that, according to the laws of combination and permutation, he would at least be one in a million, and therefore somewhat individualistic. But if he shared MOST of his beliefs with MOST others, he would be, in today's world, highly, and to me repulsively, conformist.

We should not really use the word "beliefs" here because, as the philosopher Charles L. Stevenson has brilliantly pointed out, beliefs are really statements that something is true, and in the last analysis are, if properly proposed, empirically validated. In the last analysis, scientists will all (millions of years from now) tend to have the same beliefs. Stevenson differentiates beliefs from what he calls "attitudes" but which I prefer to call "preferences" and which many others have called "values". These, he says, and I agree with him, are not empirically validatable or disprovable, and therefore it is pointless to argue about them or seek evidence for their confirmation. In the individualistic, or more individualistic world that I conceive of, beliefs would be subject to scientific verification and preferences would stand as they are. And in such a world, an individualist could share all the beliefs of others; but not, of course, their preferences. In this sense, he would be a highly social animal, and yet quite an individualist. His preferences, as I said above, also tend in OUR society to be drummed into him; but in the ideal society I conceive—call it utopian or anarchistic if you will—this would not be true, and he would be given maximum leeway to pick his own preferences.

I doubt whether we are really far apart, as usual, in our views; but language differences and the fact that we are not conversing but writing to each other probably makes us seem more so than we actually are.

Rhoda and I are still on our merrygoround, especially since I am busier than ever with patients and she is rehearsing every weekend for a dance that is scheduled to go on in January. But we see each other occasionally and manage to make the best of it. Rhoda sends her best to you and Harriet.

Cordially,

Al

Comments by Albert Ellis on His November 24, 1957 Letter

I still go along with my statement "my anarchistic bias, however, is to change the world so that we become much less a part of others and much more individualistic," but with reservations. Personally, that is still my preference, but I am not so sure today that this would be best for others. *Some* others yes. But *all?* Or even the great majority? I wonder.

Today, I am still something of an anticonformist. We are all human *persons*, but also individual people. Good!

I am amazed as I reread Rhoda's and my "Progress Report on Interdependence" that I went along with her constant socializing. I think that the amount of it led to our divorce the next year. In my thirty-five years of living with my present mate Janet L. Wolfe, our socializing is minimal and I, at least, enjoy it that way. Social chit-chat is just not my thing! I could even live without our twice-a-year institute parties (where the skits are enjoyably funny) and the yearly birthday parties the institute gives me (which fortunately take only ten or fifteen minutes). There are more important things to be done.

PROGRESS REPORT ON INTERDEPENDENCE (January 1, 1958)

We, Rhoda Winter and Albert Ellis, having stumped the experts and defied all—and we mean all—dire predictions to the contrary by remaining happi—well, married—up to and including the present writing, send

SEASONAL GREETINGS AND SALUTATIONS

to our occasional well-wishers and friends. What will happen to our unholy alliance over the next ninety years, we cannot (being logical empiricists) absolutely predict. But one will get you ten that our status remains quo for the nonce and ever after. What, you may (or may not) ask, has transpired since we drafted our infamous DECLARATION OF INTERDEPENDENCE on May 26, 1956? Read on!

*   *   *   *   *

Together, during the last annum and a half, we, the Enfants Terrible of West 56th Street, have idled along and done practically nothing except

—seen umpteen dance concerts, until they are coming out of our toes and kinesthetic souls;

—burned our candles (no symbolism intended!) at both ends and the middle;

—somehow managed to see WAITING FOR GODOT, LONG DAY'S JOURNEY INTO NIGHT, and Sexy Rexy and Droolie Julie in MY FAIR LADY;

—journeyed to Atlantic City, Long Beach Island, Folly Cove, Chicago, Urbana, Madison, New London, Philadelphia, and Brooklyn;

—enjoyed FEAR STRIKES OUT and TWELVE ANGRY MEN and viewed various other alleged movies;

—read a few good sexy novels (special list on request);

—located (!!) and furnished (!!!) a real, live, genuine, true, authentic, empirically validated apartment;

—seen 1¼ hours of TV;

—given seemingly endless parties for visiting firemen and resident denizens of the Metropolis;

—dutifully seen our respective parents, relatives, and in-laws whenever we could not find a legitimate excuse to avoid them;

—been royally entertained at parties, brunches, buffets, dinners, wakes, and sundry sordid affairs given by Dave and Elaine Berlin, Aaron Bohrod, Jo Caro and Mel Roman, Hy Chernow, Esther and LeMon Clark, Marie Coleman and Ben Nelson, Ceil and Emil Conason, Marge and Gordon Derner, Dr. and Mrs. Joe Folsom, Shirley and Skip Genther, Tore Hakannson and Walter Williams, Lila and Emanuel Hammer, Fran and Bob Harper, Morris Hartman, Dr. and Mrs. Clovis

Hirning, Mr. and Mrs. Leon Kroll, Ethel and George Lawton, Ruth and Warner Lowe, Gabrielle Lubell, Betty and Harry McNeill, Don Meyer, Alwin Nikolais and Murray Lewis, Alice Peters, Polly and Bert Pollens, Bob Sherwin, Mr. and Mrs. Raphael Soyer, Melvin Steel, Mary Louise and Lyle Stuart, Helen and Ross Thalheimer, Cora and Peter Weiss, Mr. and Mrs. Heinz Westman, the King of Siam, and the Hon. Mr. and Mrs. Addlepate Q. Addlepate;
—occasionally, when we could not help it, seen each other.

* * * * *

While all this has been going on and coming off, Rhoda has loafed along for the past year and a half doing absolutely nothing except
* assisting Margret Dietz at the Connecticut College School of the Dance for two summers and three nervous breakdowns;
* teaching and dancing with the boys and girls at the Henry Street Playhouse;
* dancing in INCREDIBLE GARDEN with and by Murray Lewis;
* dancing and mostly rehearsing in A MEDIEVAL VIGNETTE by Ronald Chase at New London;
* rehearsing and mostly dancing in RUNIC CANTO by Alwin T. Nikolais at New London;
* dancing as Walt Whitman's (believe it or not) Bride in a forthcoming dance by Billie Kirpich;
* being appointed a member of the Board of Directors of IMPULSE, the Annual of Contemporary Dance;
* becoming a full-fledged staff member and Dance Therapist at Manhattan State Hospital;
* teaching at the Hanya Holm School of Dance;
* seeing several individuals for private (and we mean private) sessions of movement therapy;
* serving as production manager for the Henry Street Playhouse Repertory Company;
* assisting Tore Hakannson and Dore Hoyer at Dore's New London and New York dance concerts;
* fighting a successful battle with the flu, three broken feet, and dancer's St. Vitus;
* giving special lectures and demonstrations on dance therapy at Connecticut College School of Dance, Overbrook County Hospital, and Minsky's Burlesque;
* rehearsing, <u>rehearsing</u>, and REHEARSING.

* * * * *

Shamed by this activity into doing something to justify his existence, Albert has given up his beachcombing days and devoted himself to various nefarious pursuits during the last eighteen months. Among other crimes, he has

+ given talks at the American Psychological Assn., the American Assn. for the Advancement of Science, the American Assn. of Marriage Counselors, the Mattachine Society, the Cooper Union Forum, the New York Society for Experimental and Clinical Hypnosis, the American Academy of Psychotherapists, and Anti-anti-vivisectionist League of Pinsk, Minsk and Kitchen Sinsk;

+ helped arrange annual meetings and workshops for the American Assn. of Marriage Counselors and the American Academy of Psychotherapists;

+ written a regular column (on you'll never guess what) for the monthly periodical, THE INDEPENDENT;

+ knocked off a dozen and a half papers for professional journals and anthologies;

+ given birth to two volumes—HOW TO LIVE WITH A NEUROTIC and SEX WITHOUT GUILT—which have just been or are about to be accouched;

+ batted out the original drafts of three new chefs-d'oeuvre—THE INTELLIGENT LIFE, RATIONAL PSYCHOTHERAPY, and EVER THE TWAIN SHALL MEET: A HANDBOOK OF HUMAN SEXUALITY;

+ helped organize two new professional groups, the American Academy of Psychotherapists and the Society for the Scientific Study of Sex and was barely dissuaded from forming the Society for the Sanctification of Sore-knuckled Sadists;

+ had infinite patience with infinite patients;

+ developed a new technique of rational psychotherapy which has aroused violent enthusiasm, 99 44/100 per cent negative, among zillions of other therapists.

* * * * *

We, Rhoda Winter and Albert Ellis, having committed all this and more since our late unlamented wedding, now, as the final notable deed in a notorious span of months, take this opportunity to wish our friends, relatives, enemies, and countrymen

A GODDAM
ENJOYABLE AND RATIONAL
NEW YEAR

ALBERT ELLIS, Ph. D.
PARC VENDOME
333 WEST 56th STREET
NEW YORK 19, N. Y.

July 6, 1958

Dear Ira and Harriet,

Good to hear from you two again. Everything in the big town is fine. Rhoda, as you well recall, had a concert in January, and it went quite well. But she had still a bigger and better one in April and that went weller and betterer. It was with one of the outstanding dancers here, Merle Marsicano, and is getting good reviews in the dance journals. She also went to Pittsburgh in June for a two-day pageant put on in the Pittsburgh Stadium by the Presbyterian Church; and that was quite a spectacle, and she enjoyed it immensely.

My book writing just keeps rolling along. Mainly, I am working on the ENCYCLOPEDIA OF SEXUAL BEHAVIOR that Nimkoff spoke to you about. It is a massive work, with authoritative articles by more than 60 authors, and entails enormous world-wide correspondence. I have been trying to figure how to get you to do a piece on it and maybe you'll have some ideas yourself. I would like you to do one on sex standards in America, but we already have a piece by Abram Kardiner on sexual morality and one on sex in America by Sidney Ditzion. Do you think one by you would overlap too much with theirs? If you can give me an idea what you might like to write on, something that is general enough for an encyclopedia, I shall be glad to have you as one of the contributors. You won't make much money on the deal, because there are so many other contributors, and each will receive $75.00 when the book is published and a proportional share of the second printing, if any. But we now have very outstanding contributors, such as Kardiner, Nimkoff, Ditzion, Robert Winch, J. P. Scott, Sandor Rado, Eustace Chesser, Abraham Stone, Oscar Riddle, Hans Selye, A. H. Clemens, Frank S. Caprio, Clara Thompson, Lester Kirkendall, Emily Mudd, Pitirim Sorokin, Robert C. Cook, Rene Guyon, Clellan S. Ford, Joseph Folsom, Morris Ernst, Winston W. Ehrmann, Lester Dearborn, Robert A. Harper, etc. So do let me know whether you are interested and what kind of a piece you think it would best for you to do.

I am returning the leaflet you enclosed. This "ENCYCLOPEDIA" is really nothing of the sort; and this is quite an old work, which has its good and bad points.

Your own doings sound somewhat fascinating. I certainly hope you are able to get a publisher for your book on sex standards. The seminar in July also seems most interesting.

I doubt whether I'll be able to get to the mountains of Tennessee next April, but if I think I can make it I'll send you along a paper. It is good to know that you intend to go to the SSSS meeting in November. The Society is going along splendidly, with lots of new members applying. As you say, it is important that it get off the ground properly, since it may prove to be quite valuable.

I certainly envy you having Harriet as a private editor. Aside from telling me how awful my stuff is AFTER it is in print, Rhoda is not the very greatest help in this connection. But then, thank heaven! I don't have to help her dance, either—aside from a few massages of her sore feet and thighs. We both plan to be in New York in November and will look forward to seeing you and Harriet then. How long do you expect to stay? Can we take you to a show—say, on Sunday afternoon, November 9th? Let us know.

Cordially,

July 10, 1958

Dear Al and Rhoda,

It was good to hear from you. We're happy that all is well with you and glad to hear about Rhoda's successes. I'm sorry we never saw her dance. When we went to the Henry Street Playhouse in April 1957 she was the director and not in the show. She did dance in these other performances you spoke of, no? I guess she must have if you had to massage her feet and thighs. Incidentally, if you must do that, then what part of Harriet's anatomy should I massage after she sits and types for me all day? Seriously we're looking forward to getting together in November with you. That Sunday afternoon (Nov. 9th) sounds fine. Is the show you mentioned one that Rhoda is in? We'd like to see that—but if it isn't then maybe we could get a cocktail and chat. We'd rather do that then just see a show, unless Rhoda was in it. Okay?

I really appreciate your asking me to contribute to your new book, ENCYCLOPEDIA OF SEXUAL BEHAVIOR. It sounds like a most worthwhile endeavor and I think there is a need for this sort of work. I would be happy to contribute to it. As you know, my own efforts for the last several years have been focused on sexual standards and thus I think I could do my best in this area. An article like the following I think would suit me fine:

<u>Current Trends in Premarital Sexual Standards</u>

Basically the article will be an examination of the integration of the integration or lack of it in our contemporary society of our four major premarital sexual standards. I will attempt to do this by showing why and how these standards are changing at the present time and what their immediate future probably entails. I will use research evidence to support my contentions whenever possible. The four standards consist of our two older ones: Abstinence and the Double Standard and our two newer "permissive" standards: one of which allows coitus only in stable affectionate relationships and the other which allow coitus regardless of the stability of the relationship. I have several concepts which I will develop here and thus this will not be just a statistical report.

The above outline ought to give you a general idea of the area I would like to write on. I do not think this would overlap with any of the other articles for I have noted that very few people discuss sexual behavior in terms of standards in an organized fashion. They usually speak of only one or two standards, if any, or at times just speak of behavior. I believe Kardiner's article on "Sexual Morality" would have more of a "moral" tone and thus would not overlap with my piece; Ditzion's work on "Sex in America" would also probably not overlap since it would likely be a historical study of sex from colonial days up to the present.

Let me know if you agree with me and if you like my suggestion. I could do the above article in a minimum of 15 pages, double spaced and I could expand it several times that amount if needed. Please tell me what audience to write for (professional, general, both?) how long to make it, when you need it, etc.etc.

Harriet and I are rushing around now getting ready for our two week visit to Chapel Hill on the 20th. We're also starting to pack, for we are moving into faculty housing in town right after we get back from Chapel Hill. I better close now and get to work. Write when you get time.

ALBERT ELLIS, Ph. D.
PARC VENDOME
333 WEST 56th STREET
NEW YORK 19, N. Y.

July 23, 1958

Dear Ira and Harriet,

A week ago I put through an article for you for the ENCYCLOPEDIA on the topic of CHANGING STANDARDS OF SEXUAL BEHAVIOR; and, so far, I have had no objections from the publisher, so it looks like it will go through all right. I had to make it broader than the one on premarital sex standards that you suggested, since Winston Ehrmann is scheduled to do an article on PREMARITAL SEX RELATIONS and yours must not overlap too much with his. But if *you emphasize changing standard, and also give some material on changing standards in adultery, petting, etc. (some of which you can get from Kinsey), your two articles should not be too overlapping. Also: the ENCYCLOPEDIA will have international distribution, and we have a good many foreign contributors; so it will be necessary for you not merely to stick to the United States (though you can mainly do so) but to say something about changing sex standards* elsewhere. *I am sure that you can easily do this.*

Enclosed are instructions for contributors, together with a sample article on frigidity, which is merely to be used for purposes of format. If you have any other questions, don't fail to ask them. If by any chance, there are objections by the publishers to your article, I will let you know. But so far, so good!

The show I mentioned in my last letter is not one that Rhoda is in, nor do I think she will be in any on November 9th. If not, cocktails and a chat will be fine with us. We'll certainly look forward to seeing both of you then.

Cordially,

ALBERT ELLIS, Ph. d.
PARC VENDOME
333 WEST 56th STREET
NEW YORK 19, N. Y.

August 6, 1958

Dear Ira and Harriet,

Oops—sorry! THIS time I AM enclosing the instructions for the article on CHANGING STANDARDS OF SEXUAL BEHAVIOR. Apparently, Hawthorn has no objection to your contribution, so you can feel safe in going ahead with the article. You may receive a formal letter of invitation from us on the article; if so, sign one copy and return it to me. If not, everything will still be OK.

Unfortunately, I cannot at the moment think of any particular sources on changing sex standards in other countries. There probably is some material of this sort in the back files of the INT. J. OF SEXOLOGY. If you have no access to these files, I could let you look at mine when you are here in November and then you could borrow any issues with relevant material.

Cordially,

Al

Comments by Albert Ellis on His August 6, 1958 Letter

I originally tried to get the SSSS started in 1950 and had a number of outstanding sexologists, including Norman Haire, Robert Latou Dickinson, and Harry Benjamin meet in New York to start it. Unfortunately, Alfred Kinsey, who wrongly thought that it would compete with his Institute for Sex Research at Indiana University, was opposed to it and killed its inception. In 1956, Hans Lehfeldt, Robert Sherwin, Henry Guze, Hugo Beigel, Harry Benjamin, and I actually got it going and by 1958 it was thriving. I am heartily grateful that, today under the name of the Society for the Scientific Study of Sexuality, it is important, powerful, and thriving. I think I can accurately say that my vigor and persistence in getting it going has borne fine fruit.

Sept. 16, 1958

Dr. Leo P. Chall, Editor
Sociological Abstracts
225 West 86 Street
New York 24, N.Y.

Dear Dr. Chall:

Ira Reiss tells me that he met you at the Danforth Seminar at Chapel Hill in July and that you told him that you would like to join the Society for the Scientific Study of Sex.

To my knowledge, I sent you an invitation, some months ago, to become a charter member of the Society; and if you are still interested, that invitations stands. Let me know in the near future whether you would like to join, since charter memberships, which obviate your making out any formal application, will soon be closed.

Cordially yours,

Albert Ellis

Dear Ira:
This should cover the invitation to Chall. Many thanks for the lead. See you and Harriet in November!

Cordially,

al

THE SOCIETY FOR THE SCIENTIFIC STUDY OF SEX

PURPOSE

The Society for the Scientific-Study of Sex (SSSS) has been organized to foster interdisciplinary exchange in the field of sexual knowledge. The aim of the Society is to bring together scientists working in the biological, medical, anthropological, psychological, sociological, and allied fields who are conducting significant sexual research or whose profession confronts them with sexual problems.

ACTIVITIES

The SSSS will hold periodic scientific meetings for the presentation of research papers. It will organize symposia, seminars, workshops, conferences, etc., to consider theoretical and practical problems in the sexual area. It will also publish a scientific journal devoted to relevant original studies and reports.

MEMBERSHIP REQUIREMENTS

Minimum requirements for Fellow: A doctor's degree or its equivalent in one of the biological or social sciences plus outstanding contributions to sexual knowledge.

Minimum requirements for Member: A graduate degree or its equivalent in one of the biological or social sciences plus contributions to sexual science; or significant contributions to sexual science.

Further information concerning the Society and its activities may be obtained from:

Robert Veit Sherwin
285 Madison Avenue
New York 17, New York

ALBERT ELLIS, Ph. D.
PARC VENDOME
333 WEST 56TH STREET
NEW YORK 19, N. Y.

Nov. 2, 1958

Dear Ira,

Many thanks for the prompt submission of your article on CHANGING SEX STANDARDS. I think that you have done an excellent objective job and that this is a valuable contribution to the ENCYCLOPEDIA.

I am happy, with you and Harriet, about the coming Blessed Event. If this is what you want, I am sure that you will derive much joy and and experiential profit thereby. Belatedly—congratulations!

Rhoda and I have just had quite a different kind of event in our married lives: a divorce. We still care for each other, and get along quite splendidly in many respects—moreso, in fact, than innumerable other married couples—but we have rationally decided that our common interests and goals in marriage aren't extensive or intensive enough to warrant our going ahead with the marital relationship. Regretful; but not, on either of our parts, disastrous. Rhoda is occupied with a trip to California, where she may possibly stay to live; and I am up to my neck in articles for the ENCYCLOPEDIA and five million other things. Damned little time even for regret.

In any event, I shall be happy to see you, if not Harriet, at the November 8th meetings. Do give my very best to the mother of your heir-to-be.

Cordially,

Albert Ellis

Comments by Albert Ellis on His November 2, 1958 Letter

As I previously noted, my "great" social life with Rhoda wasn't so great for me! Socializing, especially with dancers, was hardly my cup of tea! Rhoda and I parted amicably in November and, to this day, we are still good friends. Surprisingly, she met a physician in California and married him before the end of the year—so that they could file joint tax reports for 1958!

ALBERT ELLIS, Ph. d.
PARC VENDOME
333 WEST 56th STREET
NEW YORK 19, N. Y.

December 23, 1958

Dear Ira and Harriet,

    I think that your point about science and morals is exceptionally well taken, and that it should be a major theme at one of our regular SSSS meetings. I shall bring it to the attention of the Executive Committee and shall try to arrange for a meeting devoted mainly or exclusively to this topic. Since we already have tentatively scheduled meetings for January and March and are already at work getting speakers for the next annual meeting, I imagine that this meeting on morals and science will take place around May of 1959 or sometime next Winter. Would you like to be one of the main participants in it? If so, when would it be convenient for you to get to New York? I don't know whether anything can be arranged to coincide with your April ESS meeting.

    Yes, I met your sister, and was as favorably impressed as she. My intentions, of course, are strictly dishonorable—matrimony is something that I am not quite contemplating again at this time—but she's a fine girl and I hope to see more of her.

    Everything else is fine. I am hibernating until April, for the most part, since I loathe the cold, and don't intend to get my diabetic feet frost-bitten. But I keep busy at my usual nefarious activities—including, patients, writing, and a few professional addresses.

    The ENCYCLOPEDIA has a new editor, the famous anthologist, Whit Burnet, and he may take time to catch up with old business. So I am indefinite about the publication date, but hope to make it sometime in 1959. Most of the articles are in and most of the others promised in the near future. All told, they seem to be quite a good lot and the book should be a valuable one.

    My 1958 reprints are almost in the mail. I figure that if Harriet reads them while she is still enceinte, they may have a felicitous influence on your coming heir. Whether he'll grow up to be a sexologist or a sex fiend is not exactly clear from my zodiac readings; but what a noble experiment!

Best,

Al

ALBERT ELLIS, PH. D.
PARC VENDOME
333 WEST 56TH STREET
NEW YORK 19, N. Y.

January 4, 1959

Dear Ira and Harriet,

I will let the Executive Committee of SSSS know that you will be available some time next Fall for a presentation on values and science, and let you know what comes of the program.

No, Whit Burnet is not really an editor of the ENCYCLOPEDIA. He is the publisher's main editor, and as such will have to go over all the material in the book. I am sure that he will have some important suggestions. Your feeling that the ENCYCLOPEDIA will be an important book is certainly mirrored in my feelings, since much of the material that has come in is of a high calibre.

I am enclosing reprints of my main articles on love relationships. They are very scarce; so if you have little use for them after you have read them, please return them to me. I think they will give you a reasonably good idea of how I conducted the study. I shall be most interested in seeing your article on love when it is finished.

As ever,

Comments by Albert Ellis on His January 4, 1959 Letter

I was truly enthusiastic about the two volumes of *The Encyclopedia of Sexual Behavior* because it included definitive articles by outstanding sexologists, psychologists, and sociologists. It had articles by conservatives like Pitirim Sorokin, but was notable for liberal articles on sex, which I had mainly worked hard to gather. My coeditor, Albert Abarbanel, was helpful but actually did very little work on the *Encyclopedia*, while I kept assiduously after the contributors to do their articles. To my surprise, the *Encyclopedia* sold very well although it was expensive. For many years it exerted a profound influence in the field of sexology.

ALBERT ELLIS, Ph. D.
PARC VENDOME
333 WEST 56TH STREET
NEW YORK 19, N. Y.

February 28, 1959

Dear Ira:

We have had some trouble getting together in the executive committee of SSSS, largely because of the illness of some of the members during the last two months. We finally had a full meeting last night, however, and agreed that a scientific meeting on the subject of Sex and Science would be most desirable. We could hold this meeting sometime around next December, and would be delighted if you would be the main speaker. You can pick the specific subject you like; and also suggest some other speakers or discussants you would like to be with you on the program. Let me know your ideas in this connection, and when you think it would be most convenient for you to be in New York around December.

Meanwhile, as you will soon learn via the Newsletter, we have what should be an interesting meeting on NEED ADULTERY BE DESTRUCTIVE OF MARRIAGE on March 20th and are well on the way toward arranging the annual meeting on November 7th on PSYCHOLOGICAL FACTORS IN INFERTILITY and WHAT IS SEXUALLY NORMAL?

I trust that everything is going wonderfully well with you and Harriet. As ever, I am immersed in activities. SEX WITHOUT GUILT will soon be out in pocketbook form and I have forthcoming articles in PAGEANT, CONTROVERSEY, J. CONSULT. PSYCHOL., J. CLIN. PSYCHOL., an Italian criminological journal, etc. Also am half-finished on a new mss. on A RATIONAL APPROACH TO LOVE AND MARRIAGE, which I am doing with Bob Harper of Washington. And you?

Cordially,

March 7, 1959

Dear Al,

I was glad to hear that the Executive Committee of the SSSS decided they would have a meeting on the subject of Values and the Scientific Approach to Sex. I would be very pleased to be the main speaker at such a meeting.

Regarding the date of such a meeting, the most convenient time for me would be X-mas vacation which begins around December 20th and continues till about Jan. 5th. However, if for some reason this two-week period would not be satisfactory, I could arrange to come up early in December, or late in Nov. during thanksgiving vacation.

As for other speakers, I would like to think a little about it. How many other speakers would you want—two, three or four? There are a few sociologists such as Cy Goode of Columbia who I think would be good for such a discussion. I realize that I need not only think of people who are members of the SSSS, but I think it might be helpful if I could look over a list of members and in that way not pass up anyone who would be suitable and is a member. I take it that after I have picked these names, I send them to you and then the Executive Committee passes on them and sends out letters—is this correct? Or do I select the men and send out the letters myself?

I would appreciate some information as to format—how long should each speaker talk—is there any preference as to the sequence of speakers? Each organization seems to have its own manner of proceeding and this I want to find out what, if any, the regulations of the SSSS are concerning such a meeting. Thus, if you will inform me about this and send me a list of members, I will get started on preparing for this meeting.

That March 20th meeting on "Need Adultery Be Destructive of Marriage" sounds very interesting. Will there be any written record of this meeting? I would be interested in getting a copy of this. Do you plan to make written records of any of the meetings—I seem to remember something about this being said last Nov.

All is fine here—I am engaged in several interesting projects concerning sexual standards among negro and white high school and college students and courtship patterns among sorority and fraternity members. Harriet is getting bigger every day and should produce our heir in about 3 months.

One final thing, I am going to be on a panel discussion of sex education at the Groves Conference on April 7th. I have heard a good deal about the Groves Conferences but have never been on any of them. Do you have any helpful *information on the Groves Conferences? Dr. Stokes & Dr. Ehrmann will also be on the panel—I believe you have had some run-ins with Dr. Stokes. I've never met him—what sort of a chap is he?*

Cordially,

ALBERT ELLIS, Ph. D.
PARC VENDOME
333 WEST 56TH STREET
NEW YORK 19, N. Y.

March 23, 1959

Dear Ira,

The Executive Committee of the SSSS met last Friday, after the meeting on adultery (which was quite spirited and well attended; in fact, there was actually standing room, since we ran out of chairs), and tentatively set a date for your meeting on VALUES AND THE SCIENTIFIC APPROACH TO SEX for Friday evening, December 18, at the usual time of 8:15. We felt that putting the program on during the week of the 20th would bring it too close to the Christmas holidays; and hoped that you could get to New York by the 18th. Do you think that will be all right?

I would suggest a total of three rather than two or four speakers; or even two solid speakers, including yourself. But four gets a little unwieldly and none of the speakers has sufficient time. Normally, the Executive Committee does NOT send out invitations; but the individual who arranges the program does—which in this case is you, with carte blanche to do what you like. You do not have to check with the Committee on whom you choose; but we merely suggest that at least one fairly prominent name on the speaker list helps to draw people to the meeting.

I shall try to get you a full list of SSSS members soon. Meanwhile, I would say that some of the most relevant members you might keep in mind are Kirkendall, Murdock, Ehrmann, Tappan, John Money, and Abe Maslow. Abe would be particularly good if you could get him. As you note, non-SSSS members are perfectly satisfactory.

If there are two or three speakers, each one can have about a half hour. Sequence is up to you. The meetings usually run from 8:30 to 10:30 or 11:00.

There was no recorded record of the adultery meeting. Devereux and Clark's talks were serious and good; Ploscowe said very little in a charming way; and the minister—whose name I've already charitably forgotten—made an ass of himself, as Devereux clearly pointed out.

I don't know what I can usefully say about the Groves conference, except to say that many of the attending members, while quite well intentioned, are neither too bright nor too professionally well-informed. I have had my verbal run-ins with Walter Stokes, largely on the matter of "vaginal" orgasm, in which I think Walter still believes; but actually we are good friends, and he is one of the very few people in the country who, as in my case, will get up in public and say that premarital sex relations are fine and that there should be more of them in this country. Walter gets a little long-winded at times, even when he merely speaks from the floor; but what he has to say is usually well to the point and worth hearing.

As ever, I am, if possible, busier than usual. I have another article coming out in the June issue of Pageant; two in a new magazine called Controversy; and several others in press. Whit Burnet is now recovering from an attack of pneumonia; so the ENCYCLOPEDIA has been set back again as to publishing date; but, anyway, all the articles are not quite in yet. So things keep humming.

I gather that all goes well as Harriet approaches her Big Day. Give her, as always, my very best wishes and Noble Words of Encouragement. There's still at least one thing that men cannot do for the human race and that we'd damned better, no doubt, leave in more capable hands!

Cordially,

Comments by Albert Ellis on His March 23, 1959 Letter

Walter Stokes, a psychiatrist and sexologist, had an office in Washington for many years with my friend and collaborator Robert A. Harper. He was very liberal—as, of course, was Bob—about sexual issues, but still unduly favored "vaginal" over "clitoral" orgasms. I was the main popularizer, along with Kinsey, of the legitimacy of all kinds of women's—and men's—orgasms and had several friendly disagreements with Walter on this matter.

ALBERT ELLIS, PH. D.
PARC VENDOME
333 WEST 56TH STREET
NEW YORK 19, N. Y.

April 11, 1959

Dear Ira and Harriet:

I should normally be glad to be one of your speakers for the December 18th session on VALUES AND THE SCIENTIFIC APPROACH TO SEX, but I am not sure whether this will be advisable, since I am to be chairman of the May 22nd meeting on HOMOSEXUALITY: THE PSYCHOLOGICAL, SOCIOLOGICAL AND LEGAL ASPECTS, and probably will also chair the annual meeting program on WHAT IS SEXUALLY NORMAL? Under the circumstances, it might well not be appropriate for me to appear on the December 18th meeting, and thus to make it three in a row. Leo Chall is also to be one of the annual meeting speakers; so I wouldn't invite him, if I were you.

Also: we like to draw in as many SSSS members as possible who have not already had any main billing. So if you could get people like Abe Maslow and John Money who have not officially appeared on any of the programs yet that would be fine. I asked Robert Sherwin to send you a list of sociological members of the Society; but since his father died recently and he is in something of a mess, I wouldn't strongly count on his doing so in the near future.

From what I know of Bard it's a rather unusual school and you might well be happy there. It also has the advantage, if you think it one, of being close to the N.Y. area.

Bob Harper just wrote me saying that he had some pleasant contact with you at the Groves conference. He thought you, Pomeroy, and himself had things fairly well under control.

I can practically see your high expectations anent your forthcoming heir. I hope that Harriet, at least, keeps sufficiently cool until the finish line is reached!

Best,

Al

ALBERT ELLIS, PH. D.
PARC VENDOME
333 WEST 56TH STREET
NEW YORK 19, N. Y.

May 10, 1959

Dear Ira and Harriet:

The Bard College deal sounds fine! I think you will like it there—aside from the proximity to the Devil's Own Country, New York. Even the newborn brat will probably love it and will open its eight-week-old yap to shout Hosannas.

I agree that you should have plenty of time yet for the SSSS program in December. Some possible people you might think of, whom we have not as yet called upon for contributions to our meetings yet, are George P. Murdock, Paul Tappan, and Ashley Montagu.

Aaron Rutledge is a most reputable ex-minister who is now head of the counseling department at Merrill-Palmer and a fairly good Joe despite his clerical background. The Merrill Palmer Quarterly may not have a large circulation, but it is certainly a respectable place to publish.

You were right to tell Jessie Bernard that the ENCYCLOPEDIA will probably be out around the end of the year—or at least go to press then. Some of the contributors have been most lax about their contributions; so even though Whit Burnet is starting to get going, we must wait for these laggards.

Everything still busy as hell here. I have several new articles hot off the electric typewriter and should be at work soon on revising my sex manual, which Lyle Stuart will publish in the Fall. Never an unfrantic moment!—but I really don't take things that seriously, so manage to get through without ulcers.

As ever,

Al

Comments by Albert Ellis on His May 10, 1959 Letter

Yes, I still don't take things too seriously. So, in spite of hassles, objections, and calumnies, I go unperturbably along! At eighty-eight, still no ulcers!

May 14, 1959

Dear Al,

I've got some good news to report—no it's not the baby yet—it's my book, *Premarital Sexual Standards*, it was accepted for publication and I've signed a contract with The Free Press. Harriet and I are both very happy about it—she worked as hard on it as I did and she has been my private editor and slave.

One question has come up and I wanted to ask your advice. The editor of The Free Press wrote me and told me to write to the journals where I have published articles on the same topic as my book and to get an "assignment of copyright" from them. This I have done today. I mentioned to him about my chapter in the Encyclopedia and he said the following:

"So far as the ENCYCLOPEDIA OF SEXUAL BEHAVIOR, you will have to read your contract with them to see whether or not you agreed with them not to furnish work of competing character for another book. If you made no such agreement, you will then only have to make sure that what you have provided them does not appear in the same manner in your own book."

Now, as far as the Encyclopedia is concerned, I am not clear as to the meaning of this. I have signed no contract but I presume I will someday. Is that right? Or will they publish the Encyclopedia without contracts for the contributors? If there is a contract will it forbid me to publish material of a competing character? If it does, will that mean that my book is of a competing character?

What does he mean when he says that the material in the Encyclopedia should not appear "in the same manner in your own book"? Does he mean word for word? About half of the chapter I wrote for the Encyclopedia is on related but different topics than my book, however, th other half contains several of the overall conclusions and analysis of my book. It is not at all word for word but of course many of the same ideas are in the chapter. Is this the "same manner"?

Finally, what about an assignment of copyright—do I need to get an assignment of copyright from Hawthorn Press? The Encyclopedia will likely come out first since my book is scheduled for publication in the Spring of 1960. If I do need an assignment of copyright or if I can get such from Hawthorn please tell me how to go about it and who to write to. I guess I must need it since I needed to get it from my journal publishers—is that right? As you can see all this business about copyright and competitive books has me confused. Please tell me what you can about all this.

ALBERT ELLIS, Ph. D.
PARC VENDOME
333 WEST 56TH STREET
NEW YORK 19, N. Y.

May 17, 1959

Dear Ira and Harriet:

I am very happy to hear about the acceptance of PREMARITAL SEXUAL STANDARDS by the Free Press—which is a damned good outfit. Carol told me about it the other day on the phone; so it was no surprise to me when your letter came. Anyway, congratulations and the best of luck with the book!

As for using material which overlaps that in the ENCYCLOPEDIA OF SEXUAL BEHAVIOR, I wouldn't worry about that. The only contract you'll receive is the one you already signed with me, rather than with the publisher. As long as you note that any material in PREMARITAL SEXUAL STANDARDS which directly quotes or is fairly closely derived from your article in the ENCYCLOPEDIA comes from this article, that will be all right. In other words, simply put footnotes at one or two places on the book saying that "this material is an expanded version of material published in my article, "Changing Standards of Sexual Morality," in Albert Ellis and Albert Abarbanel (Eds.), Encyclopedia of Sexual Behavior. New York: Hawthorn Books, 1960. Or something to that effect.

As long as you have a note like this, and your book does not quote at great length from the encyclopedia article, I do not see why any assignment of copyright is necessary. If your publishers insist on it, merely write Hawthorn Books for permission to quote from your article in the ENCYCLOPEDIA, and their reply to you will be a sufficient assignment of copyright. Normal, however, this should not be necessary. Is that clear?

Everything still going most busily along here, with the interesting SSSS meeting on homosexuality coming up next week. Should be a large attendance and pertinent discussion. We shall see.

Best,

Al

July 21, 1959

Dear Al,

I'm sitting amidst a host of boxes—no, not the kind that arouse desire but the kind that create fatigue (I guess they both do!) We're moving in about 10 days and with a new baby to care for as well as pack it is difficult.

I'm starting to work on that December 18th program. I wrote to Murdock but he will be in Mexico at the time. There are several other people I want to think about and then write to two of them. Could you give me the addresses on:

    Abe Maslow
    John Money
    Ashley Montague

I'd most like to get Montague, cause he has a name which will be a drawing card. Then I'd like to get either Maslow or Money since you say they are worthwhile, particularly Maslow. Could you fill me in a little on them when you send me their addresses?

My article on love, which I wrote to you about a while back, has been accepted in <u>Marriage and Family Living</u>. I'll send you a reprint after it appears. My book is scheduled for publication this coming February—so I guess the editing will have to be done at least two months before that. It looks like my fall semester at Bard will be quite busy. Anything new on when the <u>Encyclopedia</u> will be out?

My new address at Bard College, is Annandale-on-Hudson, N.Y. We have an apartment 10 miles away in Kingston, N.Y. and the address there is 755 Broadway. If you're ever in the vicinity or feel like dropping in please do—we'd like to have you. (that means Harriet, myself, and of course, David too). He's now six weeks old and weighs over nine pounds. When he hits ten pounds—I'm telling him about girls.

ALBERT ELLIS, PH. D.
PARC VENDOME
333 WEST 56TH STREET
NEW YORK 19, N. Y.

July 25, 1959

Dear Ira, Harriet and (especially) David:

I can imagine the tribulations of your moving to the Big City (or at least close enough for comfort). It's no fun; especially if you have, as I, loads and loads of books and journals. But I'm sure you'll all make it in ship-shape order, particularly if David carries his share of the load.

The addresses you want are as follows: Ashley Montagu, Cherry Hill Road, Princeton, N.J.; A. H. Maslow, Brandeis University, Waltham 54, Mass.; John Money, Johns Hopkins University, Baltimore, Md.

I'm glad to hear that your article on love has been accepted by MFL. I have an article on the demasculinizing woman that has been accepted there, too; so we may bed together in the same issue.

Stragglers keep holding up the ENCYCLOPEDIA; and there are still about ten articles due and promised soon. I hope we can get to press late this year and off the press sometime early in 1960; but I wouldn't be surprised if the end of 1960 is more like it.

Meanwhile, I keep batting things out and have sent my new ART AND SCIENCE OF LOVE to Lyle Stuart for January publication and have another book on marriage, which I did with Bob Harper, being favorably considered, so far, by Henry Holt. So we shall see.

It will be good to see both of you again when you are around New York after you get settled. I have a trip to Lake Forest, Ill. (a psychotherapy workshop) to take next week and the APA meetings in Cincinnati in Sept. (where I appear with Hobart Mowrer and Father Curran on a wrangle about whether a sense of sin is necessary in psychotherapy); and after that I should be set for the Fall and Winter season.

October 1, 1959

Dear Al,

Well I've finally gotten a slate for the December 18th meeting on Science and Values. Just between you and me, It wasn't easy—I was turned down by A. H. Maslow, Ashley Montagu, George P. Murdock and Paul H. Gebbard. I was fortunate enough to get the acceptance of two good speakers however:
> John Money
> Walter Stokes

Counting myself that will make three speakers for the night. We plan to talk about 15–30 minutes each and then open up the floor for discussion. The basic topic will be the relation of science and values, ie., how can one objectively, scientifically treat questions of adultery, abortion, premarital coitus, etc. I envision some disagreement among the three of us and hope to get lively discussion from the floor.

Can you tell me more about my two speakers. Money is a prof. of med. pschy. and pediat.—what else? Stokes is a psychiatrist, I believe, yes? What else does he do? I'd like to know more about them so as to better handle the program. I hope to see you and them at the November meetings if I can make it. What is the date for the November SSSS?

Can you tell me more in particular about the Dec. 18th meeting. What time will it start, where will it be, what other arrangements should I make, etc?

All is fine here. Bard is a progressive sort of school with fine students and faculty—better than at W & M. I enjoy the academic life a good deal here and like the location fine—only drawback so far is that Bard is not very wealthy and the salary future is not as bright as it could be. David is a big boy now—over 15 lbs. He laughs out loud and cries out loud, thus keeping one busy and entertained. We really love him. What's new with you in N.Y.? You've probably written about 12 books since we last corresponded. I hope we can get together for a while at the SSSS meetings. Write when you have time.

ALBERT ELLIS, Ph. D.
PARC VENDOME
333 WEST 56TH STREET
NEW YORK 19, N. Y.

October 5, 1959

Dear Ira:

John Money and Walter Stokes should make two good speakers for your meeting. There should, as you say, be some disagreement and lively discussion.

The November meeting (about which you will get a program soon) will be on the 7th. Your meeting will almost certainly be held at the Barbizon Plaza, too. You had better get in touch, pronto, with Robert Veit Sherwin, 1 East 42 Street, NYC 17, to tell him definitely to arrange the Dec. 18th meeting; and with Hugh Beigel, 138 E. 94 St., NYC 28, to give him all the details, exact title, speakers and their professional positions, etc., so that he can have an announcement in the Newsletter which is to go out in November to all members and others.

I don't know any more about Money than you do and have not yet read his book but have read several of his fine papers on hermaphroditism and his article on the subject for the ENCYCLOPEDIA (which is scheduled for April 29th publication now). Walter Stokes is a psychiatrist, ex-gynecologist (I believe), marriage counselor, and holds a law degree. He is a good but long-winded speaker (make sure you hold him down to time limitations!) and has pronounced, often original views.

Things are, as ever, impossibly busy with me, what with several meetings, talks, books, articles, etc., in the near-past and near-future. I am glad that you like Bard and that the brat is becoming a big 'un. It will be good to see you (and Harriet?) on November 7th.

Oct. 31, 1959

Dear Al,

I wrote to Sherwin and Beigel on October 8th as you advised. As of now I have heard nothing—so I wrote to them again asking if all is settled. Do you know anything about this. I gave them all the information I think they needed in my October 8th letter.

I plan to be at the meeting this Saturday and will try and check with them at that time. Incidentally, if you have an extra program I'd be most grateful for a copy—I did not get any program in the mail. I'm looking forward to the meeting and hope we can get together sometime during the day.

Cordially,

ALBERT ELLIS, Ph. d.
PARC VENDOME
333 WEST 56TH STREET
NEW YORK 19, N. Y.

Nov. 3, 1959

Dr. Ira Reiss
Bard College
Annandale on Hudson, NY

Dear Ira:

I don't know why Sherwin and Beigel haven't answered you—probably because they have been too busy with the Nov. 7th meeting—but you can speak to them at the meeting.

Enclosed is an extra program. Yours, together with the monograph sent to all members, must have gone down South. You'd better make sure that the office changes your address.

It will be good seeing you again at the meeting on the 7th.

Cordially,

Al

The special luncheon, if you want it, is 7 dollars. Reservations are to be sent to Sherwin.

References

Ellis, Albert. 1957. *How to Live with a Neurotic: At Home and at Work.* North Hollywood, Calif.: Melvin Powers.

Ellis, Albert, and Albert Abarbanel, eds. 1961. *Encyclopedia of Sexual Behavior.* 2 vols. New York: Hawthorn.

Quarterly Review of Surgery, Obstetrics, and Gynecology 16, no. 4 (October–December 1959): 217–263.

Reiss, Ira L. 1960a. *Premarital Sexual Standards in America.* Glencoe, Ill.: Free Press.

———. 1960b. "Toward a Sociology of the Heterosexual Love Relationship." *Marriage and Family Living* 22 (May): 139–145.

New Books and New Institutes Blossom Forth 4
Letters from February 22, 1960 to October 4, 1967

Introduction by Ira L. Reiss

IN FEBRUARY 1960, AL informed me that he and Robert A. Harper were working to set up the Institute for Rational Living "for the purpose of teaching therapists how to do rational therapy and teaching members of the public how to live rationally." In my mind, this was probably one of the most important steps Al took in his career. It established his institute and it spurred on the entire "cognitive therapy" movement that today is so very important. Al was a pioneer and in my mind this was his most important contribution to the entire field of therapy and to sexual science. By August 1960, the institute was on its way and its first declaration of purpose and activities is included in the letters in this chapter. Today, it is a world-famous institute, now named the Albert Ellis Institute for Rational Emotive Behavior Therapy.

In my response to this first statement about his new institute, I questioned his use of the word "rational." I suggested that he seemed to be thinking of the word "rational" not just as "efficient or not self-defeating," but also as morally good. There are ways of efficiently pursing goals like abstinence that I thought Al would not see as efficient simply because he didn't like that goal. So, I felt there was a covert value judgment about rationality that I thought should be made more fully explicit. Al responded in his August 14, 1960 letter by saying that indeed he only wanted to deal with people who seek to be "unanxious and unhostile." If someone wanted the opposite, he would not tell them how to efficiently pursue it. Al saw the person pursuing abstinence as inevitably anxious or hostile. In his mind, abstinence promoted anxiety and hostility. I was not quite as certain of this as he was, but abstinence was clearly at the bottom of my list of premarital sexual standards.

In October 1960, I sent Al a copy of my book *Premarital Sexual Standards in America* and he gave me his response. He was favorable, but he thought that I was not stressing enough the arbitrariness of choosing a sex standard. He also felt that rigid sex standards, even if abided by, would produce anxiety and hostility and so were not a healthy choice. Al felt that people with such narrow standards can with therapeutic help change them and learn to accept a wider range of sexuality. It was, he felt, easier and more personally beneficial to loosen the standard and accept more sexual behavior, than to try to control the sexual behavior and accept the narrower standard. He felt I was supporting the status quo by pointing out the difficulty of changing abstinence beliefs. But I didn't think so. I thought it was a reality, especially for many young women (October 11, 1960 letter).

Al and I had discussed getting enough money to start *The Journal of Sex Research*, the official journal of the Society for the Scientific Study of Sex (SSSS). There were many expectations about when it would appear, but it took longer than we thought and the first issue was not published until 1965. As early as 1961, I wanted to be sure that I had time to make a correction in my article on science and values from the December 1959 SSSS meeting. I had written in that article about Diana being born from the head of Zeus and of course it should have been Athena who had that unusual birth passage. I knew I couldn't rewrite Greek mythology. As it turned out, I made the correction, and my article appeared in a book of SSSS papers that came out two years before the journal (*Advances in Sex Research* 1963).

Al asked me to be on his advisory board for his new institute. But at that very time I had become disenchanted with the anarchy at Bard College and had just accepted an offer to go to the University of Iowa in Iowa City. The University of Iowa was a big-ten college and it had the research expertise that I needed to carry out my NIMH research grant. So I gave up on small liberal arts colleges and their freedom to lead a chaotic existence and settled for a major university and its promotion of research and publication. I wonder which choice Al would have made given his greater proclivity for anarchy. But I wanted to do research and publication at that point in my career and so the move to Iowa fit in quite well.

In his June 4, 1962 letter, Al wrote about the Midwest and particularly about his trip to Minneapolis. He noted: "I find psychologists in the Midwest more congenial to my approach than in most other sections of the country. I think they're saner and more scientific." That was one opinion we had no argument about—I stayed at Iowa eight years until I left to go to the University of Minnesota in 1969. I was to stay there for the rest of my academic career. Al told me of his new books and I shared with him our anticipation of our second child, scheduled to arrive in July 1962. Of course, I also had other publications but Al was so much more prolific than I was that I didn't even think of comparing numbers with him.

After 1962, our correspondence dwindled down but we did see each other at professional meetings as documented in the 1965 letters in this chapter. Then in September 1967 I sent him a copy of my second book *The Social Context of Premarital Sexual Permissiveness*. Al always was supportive of my work and he praised this book too. But he was not averse to making critical comments and he disagreed with my assertion of equality in innate sexual desires that I put forth in this book, and wrote in his October 4, 1967 letter: "I would wonder . . . whether the innate sexual drives, of our males and females, really are that relatively equal." He also raised the possibility that "the female of the species may have an innate drive toward conservatism that is rather greater than that of the male." To be sure, he wasn't asserting these views but rather raising the issues more than I did. But to me it further illustrated his greater reliance on biology at this time.

Regardless of such differences, we agreed on far more than we ever disagreed about. We had an essential agreement on the great value of reason and science and the value of being flexible and skeptical in one's professional approach. We differed, I believe, on the specific "weights" we would give each of our personal values, but we had the same fundamental values. Besides science, we both endorsed pluralism, gender equality, freedom of choice, and, of course, the importance of sexual behavior. I think our letters in this book afford a clear insight into our personalities and our values in those early years. They also afford a portrait of our life at the very time that the country was igniting the fuse of the sexual revolution. We were trying to think through the key sexuality issues that people were facing and we were both trying to help people understand what sexuality was all about so they could make choices that best suited them.

ALBERT ELLIS, PH. D.
PARC VENDOME
333 WEST 56TH STREET
NEW YORK 19, N. Y.

Feb. 22, 1960

Dr. Ira Reiss
Bard College
Annandale On Hudson, NY

Dear Ira:

No, the June publication date is not going to be kept by the ENCYCLOPEDIA. Late this year or early next year is more like it. Galley proofs will eventually be sent you—but not for months, probably. You will get a free copy of the book, and, I hope, some reprints of your article.

I am sending you under separate cover a copy of my monograph on Freudianism, which gives some perinent references. Also you might look up H. J. Eysenck's penguin book, THE USES AND ABUSES OF PSYCHOLOGY.

The Journal of Sexual Research is progressing; but no actual financial support for it has yet been forthcoming; and we are still in need of this. Your article will be published if and when the journal gets going.

My new book, THE ART AND SCIENCE OF LOVE, will be out soon and I shall send you a copy in a few weeks. My newer one, with Bob Harper, CREATIVE MARRIAGE, will not be out till the end of this year or early next year.

When you make changes on galleys you should, as much as possible, use official proofreading marks. If not, make things as clear as possible to the printer.

I am glad to hear that Harriet and David are well and that Harriet still has enough time to help you with your book. I barely have enough time to help myself with my own books! Have been almost hibernating for the cold months, working like hell on innumerable projects, including setting up the INSTITUTE FOR RATIONAL LIVING, INC., which Bob Harper and I are starting for the purpose of teaching therapists how to do rational therapy and teaching members of the public how to live rationally. Quite a major project it may turn out to be. I shall let you know how it keeps coming.

Cordial regards,

Comments by Albert Ellis on His February 22, 1960 Letter

Actually, the Institute for Rational Living—later to become the Albert Ellis Institute—was started in 1959 by me, my oldest friend Manny Birnbaum, and Michael Feinstein, a lawyer and an accountant. Bob Harper was always an advisor but not too closely allied with it, because he kept living and working in Washington, D.C. However, he generously gave all his share of the royalties on *A Guide to Rational Living* (1961) and *Creative Marriage* (1961) to the Institute and has always been a prime supporter of it and its work.

March. 1, 1960

Dear Ira,

Here is a copy of my latest book, THE ART AND SCIENCE OF LOVE, which Lyle Stuart is publishing on March 21. I should greatly appreciate your noting, as you read it, any typographical, grammatical, factual, or other errors that you may find and letting me have a list of these (with page numbers) so that future editions of the book may be corrected. Also, if you have any ideas about significant material or references that might be added, please let me hear about them.

Most cordially,

Albert Ellis

333 West 56 Street
New York 19, New York

ALBERT ELLIS, Ph. D.
PARC VENDOME
333 WEST 56TH STREET
NEW YORK 19, N. Y.

July 29, 1960

Dr. Ira Reiss
Bard College
Annandale-On-Hudson, NY

Dear Ira:

Many thanks for the reprint of your latest paper on love. I am particularly glad to have it since my May issue of MFL got lost in the shuffle and has not reached me yet.

Did you get the copy of my new book, THE ART AND SCIENCE OF LOVE, that I sent you in March? I vaguely recall that you acknowledged receipt of it, but am not sure. If you did not get it, let me know and I shall trace it. I have a lot of other things recently off the press and shall collect them and send them, as usual, by the end of the year.

I am happy to hear that your book will be out soon. You can add it to the references in your ENCYCLOPEDIA article, which you will be receiving proofs on any day now.

If I get to the ASA meetings at the Statler, I shall be delighted to see you and Harriet. I may never get there, largely because the Amer. Psychol. Assn. meets, starting the middle of that week, in Chicago, and I have to get there for a symposium in which I am participating. Anyway, David sounds like a great chewer off the old block; so give him a wad of gum for me.

Your love paper hits several things right on the head and I am sure it will enhance your reputation. Your wheel theory has some excellent points to it. I am not entirely enthused about the direction of your arrow, since it gives the impression of the process's being less interactional than does your text; and I don't know if it emphasizes sufficiently the fact that rapport frequently CONSCIOUSLY stems from some of the other elements in the wheel, especially personality need fulfillment. I also feel that your pilot study may have neglected some other important factors, such as the essential beauty, charm, grace, etc. of the chosen partner. Have you read Vernon Grant's PSYCHOLOGY OF SEXUAL EMOTION? He has some excellent observations to make.

I am enclosing a tentative prospectus on the new Institute for Rational Living, Inc., which is in the process of organizing. If you have any comments, by all means return one copy to me and keep the other for your files. And if you are interested in being on the Advisory Board, we shall be most pleased to have you.

Cordial regards,

Albert Ellis, Ph.D., Exec. Director
Institute for Rational Living, Inc.

The Institute for Rational Living, Inc.
333 West 56 Street
New York 19, N.Y.

The Institute for Rational Living, Inc. has been organized for professional training and research in the principles and practice of rational psychotherapy and for public education in rational living.

Professional Research and Training
One of the primary purposes of the Institute is to serve as a research and training center where professional workers—including psychiatrists, psychologists, psychiatric social workers, marriage counselors, and other workers in the field of human relations—can study and investigate the most effective methods of psychotherapy and counseling. To this end the professional research and training program of the Institute will include the following functions:

1. The Institute will establish a full-time training program for psychotherapists and give a certificate in psychotherapy to qualified professionals who complete a two or three year program in the theory and practice of rational therapy.

2. It will give special courses in psychotherapy, marriage and family counseling, sex education and therapy, the therapeutic education of parents and children, rational group psychotherapy, and similar specialties to qualified professionals who are not able to become full-time matriculated students of the Institute but who are interested in certain aspects of training.

3. It will conduct short term workshops, seminars, discussion series, lectures, and meetings for the training of professional psychotherapists and counselors as well as for members of adjunct professions, including educators, sociologists, anthropologists, and vocational counselors.

4. It will establish and operate a full-time psychological clinic for research and training purposes.

5. It will sponsor and operate a thoroughgoing research program in relation to the theory and practice of psychotherapy, and will particularly concentrate on research into the relative efficacy of different kinds of psychotherapeutic procedures.

6. It will maintain a specialized library of books, articles, recordings, and films for the use of the staff and students of the Institute. Recordings and films of therapeutic sessions will also be available, for purposes of education and training, to qualified professionals, clinics, and hospitals that are not affiliated with the Institute.

7. It will maintain a speaker's bureau which will arrange for members of the staff of the Institute to participate in outside lectures, discussions, and other meetings on psychotherapy and human personality, and to present at these gatherings the most pertinent findings of the Institute relating to effective psychotherapeutic techniques.

8. It will publish a journal, RATIONAL LIVING, which will include articles by the professional staff of the Institute giving the latest findings of the research sponsored by the Institute. The journal will also include relevant articles, reviews, and abstracts on psychotherapy and rational living by writers and professionals not connected with the Institute.

<u>Public Education and Service</u>

A primary aim of the Institute is to carry on extensive and intensive education and psychotherapeutic work with members of the public, and specifically to help as many people as possible to discover and be able to apply rational methods of understanding and overcoming their life problems and their emotional disturbances. To this end the public education and service program of the Institute for Rational Living, Inc. will include the following functions:

1. It will conduct a regular and special series of lectures, discussions, seminars, film showings, and meetings for interested members of the general public. This public education program will include presentations on personal problems, marriage and family relations, sex education, and other aspects of modern living.

2. It will establish and operate a full-time clinic where members of the public, on a low-cost basis, may obtain intensive psychotherapy and counseling on both an individual and group therapy basis.

3. It will establish and maintain special services for disturbed individuals who are residing in the community or in psychiatric institutions and who are not in a position to attend the Institute's clinic or who may benefit by a less formal therapeutic approach.

4. It will, as soon as is feasible, establish and operate a full-time grade school where students, in addition to learning the normal educational curriculum, can learn from an early age to think clearly and rationally about themselves—can learn preventively and therapeutically to tackle their minor and serious life difficulties and thereby help themselves become healthier and happier human beings.

5. It will provide speakers from its staff to outside organizations and institutions, so that the principles and practice of rational living may be more effectively taught to members of the general public.

6. It will publish a journal, RATIONAL LIVING, which in addition to its professional functions (mentioned above) will include articles and abstracts designed to help the intelligent layman in accordance with the principles of rational psychotherapy and rational living.

7. It will publish and distribute pamphlets, monographs, and books which, like the articles in the Journal of the Institute, are designed to help members of the public understand the principles of rational therapy and rational living.

<u>Advisory Board</u>
Joseph Andriola, M.S.W., Ph.D., Atascadero State Hospital, California
Emmet A. Betts, Ph.D., Betts Reading Clinic, Haverford, Pennsylvania
Hugo G. Beigel, Ph.D., Long Island University
Roger J. Callahan, Ph.D., Eastern Michigan College
LeMon Clark, M.D., Fayetteville, Arkansas
Anna Kleegman Daniels, M.D., New York City
Lester W. Dearborn, Counseling Service, Boston, Massachusetts
Gordon Derner, Ph.D., Adelphi College
George Dolger, Ph.D., New York City
Henry Guze, Ph.D., Long Island University
Robert A. Harper, Ph.D., Washington, D.C.
Frances R. Harper, Ed.D., Arlington Board of Education
Edward J. Humphreys, M.D., Division of Behavioral Problems, Department of Health, Commonwealth of Pennsylvania
John W. Hudson, Ph.D., Merrill-Palmer School
Sophia J. Kleegman, M.D., New York University-Bellevue Medical Center
Herman R. Lantz, Ph.D., Southern Illinois University
Lucile Lord-Heinstein, M.D., Alston, Massachusetts
Nathan Masor, M.D., Staten Island, New York
Charles L. Odom, Ph.D., New Orleans, Louisiana
Harriet F. Pilpel, LL.B., New York City
Murray Russell, M.D., Garden Grove, California
Robert Veit Sherwin, L.L.B., New York City

<u>Inquiries</u>
For further information write to: Albert Ellis, Ph.D., Executive Director, Institute for Rational Living, Inc., 333 West 56 Street, New York 19, N.Y.

Comments by Albert Ellis on His July 29, 1960 Letter

The Institute for Rational Living was, as I have noted, already set up in 1959 but was not yet accepted as a not-for-profit institute by the Internal Revenue Service. It started to carry out the projects mentioned in this prospectus, was chartered by the Board of Regents of the State of New York in 1968, and has thrived as a training institute for counselors and therapists and as a public education institute until the present day. Its psychological clinic sees hundreds of clients every year, it gives public and professional workshops all over the world, it has over twenty affiliated institutes and training centers in the United States and abroad, and it publishes and distributes many pamphlets, books, audiovisual cassettes, and other materials on rational living for the profession and the public.

ALBERT ELLIS, PH. D.
PARC VENDOME
333 WEST 56TH STREET
NEW YORK 19, N. Y.

August 5, 1960

Dr. Ira Reiss
Bard College
Annandale-On-Hudson, NY

Dear Ira:

Many thanks for your comments on the prospectus for the Institute for Rational Living, Inc. They will prove helpful in revising it.

You are right about the 18th century rationalists—who irrationally believed that rationality was an absolute value. As Bob Harper and I explain in our latest work, RATIONAL LIVING IN AN IRRATIONAL WORLD, we merely use the term "rational" to mean "efficient," "non-self-defeating," "logically consistent," etc.; and we do not think that rationality is an absolute good.

The principles of rational living are best exposited in this book, which probably won't be published for about another year. I'm afraid you'll have to wait until then for a fairly good exposition; and for an even better one in our book, RATIONAL PSYCHOTHERAPY, which will not be published perhaps for two or three years.

Cordial regards,

Comments by Albert Ellis on His August 5, 1960 Letter

Actually, *Rational Living in an Irrational World* was published as *A Guide to Rational Living* in 1961 and has gone through three revisions and sold almost two million copies since that time. Not to mention, many foreign translations, which again have sold hundreds of thousands of copies. *Rational Psychotherapy* was accepted for publication by Paul Meehl but was refused by the publisher for no good reason as far as I could see and finally got published in 1962 as *Reason and Emotion in Psychotherapy*.

August 9, 1960

Dear Al,

I received my Galley Proofs from the Encyclopedia yesterday and made my corrections and mailed them back today. I liked the editorial work—did you do that or someone else? It has been almost two years since I wrote that article in October of 1958 but with a little revision it sounded good to me. I'm looking forward to the publication of the Encyclopedia in Feb. 1961. There are several articles in there that I want to read myself.

One thing that has to be changed is my biography. It is out of date as I wrote it 2 yrs. ago. I have enclosed two copies of a new biography and I'd appreciate it if you would use this new one in the book.

I'm very much interested in your concepts of "Rational Therapy" but I still get the feeling that the word "Rational" is being used in more ways than one. For example your definition of in your last letter was "efficient", "non self-defeating" and "logically consistent." It seems to me that these three definitions are not at all the same thing. A particular action or way of thinking can be quite efficient FOR CERTAIN ENDS and quite inefficient FOR CERTAIN OTHER ENDS. Again, a pattern of thinking can be efficient and yet be "self defeating" in that by getting you one thing it thereby denies some other part of your life. Also a pattern can be logically consistent like "if you think you are worthless then you ought to treat yourself in a worthless fashion," and yet I feel sure that this type of thinking could also be called "self defeating" and not very "efficient" for certain ends. Thus, I still contend that these 3 characteristics do not define a single entity and actually define three separate types of things.

Secondly, I would still suggest that the word "rational" as you seem to be using it, contains a covert value judgment. I say this because if used objectively an act of suicide could be called rational if all you mean is efficient means to an end (that is, if it was carried out in the most efficient manner). In short, when used to mean efficient means to an end rational can be applied to all sorts of actions which I am sure you would not want to apply it to. I think what you mean by "rational" is that it is the kind of action and thinking which you MORALLY EVALUATE AS GOOD. In short, I think you select from the total range of efficient actions those which you like, which you prefer. To illustrate: What advice would you give someone who believed strongly in abstinence before marriage but was having difficulty of some sort in living up to his belief? How would you apply your rational principle to such a specific case? Would you see which path of action would likely be most efficient in achieving HIS GOALS or would you try to "liberalize" his goals so as to make them closer to what YOU ACCEPT. Would you actually look to see what was the most efficient or less self-defeating or what have you behavior or would you proceed on the premise that you know what is the most efficient behavior and instead of checking empirically your notions in this particular case, proceed with your general notions of "efficiency." My point here is that unless you carefully and objectively define what is meant by rational you can all too easily use it as a synonym for "what I think is best." I may be all wet but I get the feeling from your three-part definition that you are using the term this latter way. I'll be interested in your comments.

Lest you assume from the above that I don't like your project, let me quickly add that I think (as I believe I told you before) it's a most ambitious and admirable endeavor. It's just that when you use phrases like "rational living" I get the impression that this is not a scientific enterprise but an ethical one for it is from ethics that we get specific and general rules for living, not from science. All we can get form science are the types of consequences that follow from certain actions. The choice among these as to how to live is an ethical one that science can at best only help us with. Since there are many rational, non self defeating and efficient ways to live—how does one choose among them?

Best wishes,

ALBERT ELLIS, Ph. D.
PARC VENDOME
333 WEST 56TH STREET
NEW YORK 19, N. Y.

August 14, 1960

Dr. Ira Reiss
Dept of Sociology
Bard College
Annandale On Hudson, NY

Dear Ira:

It may be too late to change your biography in the ENCYCLOPEDIA, as I already corrected the page proofs on the front pages a couple of weeks ago. The one that is now in is not substantially different from the new one you sent; and the material on your Public Health Service research grant is not too appropriate, anyway, since we did not include similar material on the other contributors. The present biography does include your new position.

You are of course right in pointing out that rational therapy has some underlying value judgements. These are, as shall be pointed out in detail in the introductory chapters of our book on the subject, that living with intense, prolonged, and frequent anxiety and hostility is undesirable; and living without these kind of feelings is desirable. Anyone who believes that living in an intensely and enduring state of anxiety or hostility is desirable is of course entitled to his belief; and if so, as will be noted, he will be wasting his time trying to be "rational" in the sense we employ the word. If he WANTS to be unanxious and unhostile, then we can tell him —and we hope scientifically rather than evaluationally-how to be so in an efficient, non self-defeating, and logically consistent way.

Thus, the abstinent individual you ask about would be efficiently and self-consistently ABSTINENT; but it would be our contention that he would be, because of his abstinence, in most instances (consciously or unconsciously) anxious and/or hostile. If he wants to be abstinent and disturbed, that is his privilege; but if he wants to be unanxious and unhostile, he damned well better not be abstinent. Not because we think that abstinence per se is "best"; but because we think (and perhaps wrongly) that abstinence, or the attitudes he would usually employ to remain abstinent, are incompatible with anxiety and/or hostility.

I am sure that this brief reply will not answer all your questions; but when the introductory chapters to RATIONAL PSYCHOTHERAPY are written I shall be most happy to send you carbons, so that we can see (mostly for our benefit) whether we are really being as unconsciously evaluative as you think.

Cordial regards,

Comments by Albert Ellis on His August 14, 1960 Letter

As I noted in my reply to Ira, I answered his questions about rationality briefly and not too adequately. A fuller answer would be: Humans, individually and as a group, start with desires and goals (a) to stay alive, (b) to relate to other people, (c) to relate to a few people intimately, (d) to get information and education to aid their goals, (e) to work productively and earn a living, (f) to be safe from pain and harm, and (g) to enjoy various recreations. In regard to these goals and desires—as Ira says—I am using "rational" to mean: "What I think is best." People's goals and desires are not fully empirically, scientifically, or objectively chosen, except that if they had no desires, they could hardly survive.

They largely *choose* to feel healthily frustrated, sorry, and disappointed or unhealthily panicked, depressed, and enraged when their desires are thwarted and/or they get what they don't want. Seeking fulfillment of their desires to some extent enables them to stay alive and "happy." Choosing to feel panicking, depressing, and raging sometimes helps—but often sabotages them.

If they desire to attain their goals—which they of course do not *have to* do—they can probably choose certain thoughts, feelings, and behaviors rather than others; and if they do so, we can call their action—as rational emotive behavior therapy (REBT) tends to do—sensible, rational, efficient, self-aiding, or productive. But as Ira points out, their self-helping choices are far from universal and are in some respects value judgments. Different strokes for different folks! So no behaviors are *completely* "rational" or "irrational" for all people under all conditions.

Nonetheless, if you define—yes, define—your goals as "desirable" or "good" and you define other goals as "undesirable" or "bad," you can *somewhat* precisely and "scientifically" discover which of your thoughts, feelings, and actions produce more "good" than "bad" results, under certain conditions! So REBT holds that A (Adversities, Activating Conditions) times B (Beliefs-Feelings-Behaviors) lead to C (emotional and behavioral consequences or results).

"Rational," as used in REBT, essentially means behavior that (usually, not always) leads to chosen "desirable" consequences. It is never an absolutistic term, but it *often* means efficient, helpful, desirable, non-self-defeating thinking-feeling-behaving. But all these terms *themselves* are never perfectly clear and absolutistic. Otherwise, REBT would be what it opposes: rigid, inflexible, and absolutistic! Like science, it seeks—but never *demands*—openness and flexibility. If we agree on our goals, we presumably can "rationally" and "empirically" discover whether we actually achieve them—and whether their achievement gives us "good" results. Often, our most desired "rational" goals turn out to be, when we achieve them, what we really *don't* want. If so, we can always revise them for the future.

ALBERT ELLIS, Ph. D.
PARC VENDOME
333 WEST 56TH STREET
NEW YORK 19, N. Y.

October 7, 1960

Dr. Ira Reiss
785 Broadway
Kingston, NY

Dear Ira:

Many thanks for sending me a copy of PREMARITAL SEXUAL STANDARDS IN AMERICA. I think that you have got by far the most out of what could otherwise be a limited topic and that you have done [a most excellent, remarkably objective job] of delineating the standards and discussing them. I note that Dick Ehrmann's book has just come out in pocketbook form; and I see no reason why yours should not be able to get a wide distribution, later on, in this kind of format. It is well written and quite interesting and should have more than a professional reader interest.

Since you have been so [unusually objective] there is virtually nothing for me to quibble with in your book. The only point I might take some issue with is your at times super-objectivity, which is fine for the purpose of this particular book, but which I doubt that you always personally hold. Thus, on page 209 you note that "The serious situation is the one involving a wide gap between desire and standard. The solution to conflict of this kind is easy to state but difficult to achieve. One must either change his desires or change his standards—or change both so as to bring them closer together, thereby lessening the intensity of the conflict. . . . The answer is in lessening the gap between desire and standard, and this can be accomplished either by indulgence or chastity depending on what the individual can accept and achieve. Some people feel so strongly about remaining chaste that the only workable solution is to try to lessen their sexual desires by avoiding certain kinds of situations."

The implication of this statement is that sex standards are fairly objectively and realistically chosen in the first place; and that even if they are unrealistically chosen, it is virtually impossible for an individual with high chastity standards and high sex drives to change his standards. These implications are questionable. What is more, they mask value judgments on the part of those that make them—as you are normally the first to point out. In the first place, sex standards (as you show in *other sections of your book) are hardly objectively chosen; nor are they often realistic. One of the main characteristics of our own sex standards is that they are very <u>arbitrary</u>; and that because of this fact they are often unrealistic. You emphasize this in the first part of your book; but fail to re-emphasize it, I think, in this section on the possibility of changing one's standards.*

Secondly, even when sex standards are arbitrary, unrealistic, and rigid they definitely <u>can</u> be changed. The human animal is the one animal which can change almost any of its views, with or without psychotherapy. And with enough time and effort, preferably helped by a trained therapist, a highly sexed chastity-biased individual can definitely change his "ideals" no matter how strongly he feels about them. Every day in the week, almost, I help such strongly biased individuals to see how arbitrary their standards are and to change them. Your statement, therefore, that "others are unable to change either desires or standards and end up with strong emotional disturbances" is much less true of standards than desires. They find it <u>difficult</u> to change their standards; but they can, unless they are hopelessly psychotic, almost always do so.

Your "objectivity" in this particular section (as in a few but not many others) is therefore questionable. You are really to some extent upholding the present antisexual status quo by implying that it cannot be changed, when you merely should be pointing out, I think, that it is difficult to change it. Is this not so?

In any event, I think your book is a fine contribution to a very muddled field; and I do not by any means place it in any reactionary class (as I would place the work of Robert O. Blood and many others). I wish you great success with the book and with your present and subsequent efforts.

As ever, please give my very best to Harriet. I am sure that she is enjoying your giving birth to this work almost as much as she enjoyed, awhile ago, bearing The Brat.

Cordial regards,

Comments by Albert Ellis on His October 7, 1960 Letter

I say in my letter to Ira that "one of the main characteristics of our own sex standards is that they are very arbitrary." Not exactly! They are set in the light of distinctly selected *goals*, and for these goals, they are hardly arbitrary. Thus, Judeo-Christian sex goals are, primarily, to preserve marriage and the family—and sometimes at all costs. These goals make some definite sense, but they are (somewhat arbitrarily) chosen instead of the goals of putting sex pleasure and liberalism *over* family preservation. The goal of family preservation is not entirely arbitrary, since it may—who knows?—aid human survival more than, say, the goal of promiscuous sex or the goal of human abstinence until people reach the age of thirty.

I note that our sex standards "are often unrealistic." Yes, the Judeo-Christian standards go against some biosocial processes, such as human urges for immediate sex satisfaction, for adventure, and for new experiences. But Jews and Christians have nonetheless largely abided by them for thousands of years—and are still alive and kicking!

My statement that people "find it difficult to change their standards; but they can, unless they are hopelessly psychotic, almost always do that," is, of course, exaggerated. Individuals who are psychotic or suffer from severe personality disorders often perpetuate rigid sex (and other) standards, but they sometimes *too easily* change from strict to liberal standards, or vice versa.

October 11, 1960

Dr. Albert Ellis
333 W. 56 St.
N.Y. 19, N.Y.

Dear Al,

Thanks very much for your kind words about my book. I am very happy that you enjoyed reading it. I don't know how the book is selling in general but it's selling quite well here at Bard. The college kids seem to get quite a bit out of it and I'm quite glad about it. I believe my publishers sent a copy to Beigel to review for the Newsletter and to Sherwin to get comments on officially from the SSSS. There will, of course, be other reviews. <u>Coronet</u> will do an article on it in their December issue. Sales are nice, but I'm most concerned with the professional reaction to the book and that's why I'm glad you liked it.

As for your comments about my statement on desire and standards—I think I would mostly agree with you. that is, it is an empirical question as to whether it is easier to change standards than desire. I would generally agree that it is from my own knowledge easier to change standards. But—I can't rule out the possibility that in "some" people, and that is how I put it, it may be easier the other way due to very rigid training and experience. I admit that here too, this is an empirical question, and perhaps in actuality there are no such cases, or at least extremely few. Frankly, I don's know. You say you can almost always change standards but then is that conclusive evidence since you do not try to change desire in your patients? Would we not have to check with a very conservative therapist and see if his attempts to change desire and maintain abstinence standards are successful? In short, I felt that since I lacked evidence on the point of changing desires versus changing standards, I would be fairest if I pointed out that both can be changed and it depends on the person as to which is easiest. Further it seems that even if almost all therapy patients found it easier to change standards rather than desire, we would not have a conclusive answer because such patients do not necessarily represent all people with such conflicts and those that do not go to therapists may well resolve their conflicts by controlling desire—no? In the same vien, I felt that some people might get caught between equal pressures from standards and desires and might not go for or get help from outside and in such cases emotional disturbances can mount. Again, I am willing to test this empirically and see actually how many people are "stuck" in such dilemmas.

In addition, you know of course that I have no "masked values" in the direction of favoring Abstinence so that was not the purpose of my comments on standards and values. I just honestly don't know if it almost always is easier to change standards. If there is good evidence on this I would be most interested and I guarantee you I will incorporate it in any revision of my book that would be made. This is an area where the evidence would be more in your field than mine and I'd appreciate any information on this you can give me.

I liked your comments on the above section cause I do want to have such analysis of the book. If you have any other comments, pro or con, I'd appreciate hearing them.

We plan to be at the SSSS convention on the 5th in N.Y. and look forward to seeing you there.

Cordially,

ALBERT ELLIS, Ph. D.
  PARC VENDOME
333 WEST 56TH STREET
  NEW YORK 19, N. Y.

Dec. 20, 1960

Dr. Ira Reiss
College of William & Mary
Williamsburg, Va.

Dear Ira:

At the last annual meeting of the Society for the Scientific Study of Sex, it was suggested by one of our members that since we are having financial difficulties in starting the JOURNAL OF SEXUAL RESEARCH which we intend to publish as soon as possible, we might get together three thousand dollars, which would easily see us through our first year of publishing the journal, by borrowing fifty dollars from each of sixty members of the Society. If the journal gets through this first year, it is most likely to be self-supporting thereafter.

This suggestion was enthusiastically received by most of the members present at this annual meeting and was later endorsed by the Executive Council of the Society. Accordingly, I would like to ask if you would be one of the members of the Society who would lend it fifty dollars, so that we would quickly get the journal started. If so, please send a check, with a note saying that you are lending this amount to the Society for the purposes of starting a journal to the Executive Secretary, Robert Sherwin, Esq., 1 East 42 Street, New York 17, N.Y.

Cordially yours

Albert Ellis

March 12, 1961

Dear Al,

Just a brief note to let you know what's new. First, congratulations on your new book, CREATIVE MARRIAGE. I saw it advertised in the N.Y. Times today. How you get the time to do so much writing I'll never know. Best of luck with the book. I was glad to see that the N.Y. Times does advertise your publications now—I remember you mentioning that for a time they would not do that.

The news with us is that I have accepted an excellent offer from the University of Iowa. The position is at a higher rank, several thousand more in pay, and full charge of developing a graduate program in the sociology of the family. Iowa is one of the "big ten" schools and has a relatively large sociology dept. and a Ph.D. program. The chance of working with graduate students and especially in my specialized area of the family is what interests me the most. We expect to go out there the end of August, so I'm sure we'll see you before then. I'm giving a paper on April 8th at the Eastern S.S. meetings at the Sheraton-Atlantic Hotel in N.Y. If you are at those meetings we can get together.

My book seems to be going quite well according to what the publisher tells me—he said it had "healthy" sales and was one of his "best-sellers." I'm not sure just what that means, but I'm not expecting to be a millionaire.

Write when you get time and let us know what's new with you.

Best Wishes

ALBERT ELLIS, PH. D.
PARC VENDOME
333 WEST 56TH STREET
NEW YORK 19, N. Y.

March 18, 1961

Dr. Ira Reiss
Bard College
Annandale-on-Hudson, NY

Dear Ira:

Thank you for your comments on CREATIVE MARRIAGE. I am getting a copy off to you, but will not be able to get around to mailing it until after I return from the Midwest trip I am about to take. Several places, including the Univ. of Minn. and the State Univ of Iowa, have asked me to speak; so I am getting them all in next week. It should be fun.

I am glad to hear that you are going to Iowa next year. It certainly is one of the big ten. I have never been there myself, but shall be speaking there in a few days. Leonard Goodstein of the Dept. of Psychology and head of their Counseling service is an old friend of mine; and I have known other people there. I am sure that you will like it.

I shall certainly try to see you and Harriet at the ESS meetings in April, but am not sure whether I shall be able to make it. I hope that, one way or the other, we can get together before you leave for what to New Yorkers seems to be the Wild West.

You should have got your copy of the ENCYCLOPEDIA by now; if not, ask Hawthorn what happened to it. You should also be receiving a check from them soon for your first royalty payment.

My very best, as usual, to Harriet and the Brat.

Cordial regards,

June 19, 1961

Dear Al,

Just a note to say thanks very much for sending us a copy of your new book, A GUIDE TO RATIONAL LIVING. School is just finishing up here so I haven't yet had a chance to read it but I am most interested in reading it as I've heard so much from you about your approach and am anxious to see how you describe it in more detail. Thanks also for the pocketbook version of THE FOLKLORE OF SEX.

We hope we can get to see you this summer before we leave to go to the University of Iowa.

Best Wishes,

Ira L. Reiss

June 26, 1961

Dear Al,

In reading over the June issue of the NEWSLETTER I noticed that it now seems assured that the JOURNAL OF SEXUAL RESEARCH will appear in March of 1962. My congratulations on another achievement of note. I think there is a real need for such a journal. If you recall, I gave a paper in December 1959 at the SSSS meetings entitled: "Personal Values and the Scientific Study of Sex." I sent, on request a copy of this paper to Ruth Doorbar. You wrote to me shortly after saying that the paper would appear in the JOURNAL OF SEXUAL RESEARCH whenever it came out. My reason for writing is just to be sure one point is changed before the paper is published. On p. 9 I speak of the Goddess Diana being born full blown from the head of Zeus. Well, I've got the wrong girl—it should be the Goddess ATHENA. I suppose there will be galley proofs but just in case I wanted to mention this error of mine.

I hope all is going well with you. I'm looking forward to reading your new book on rational therapy—my sister read it and said she liked it a great deal. I'm in process now of writing an article for the ANNALS on sexual codes of teenagers—it is to be in the November 1961 issue. It probably takes me as much time to do this article as it does for you to do a book or so. This year in particular you've been turning out books at a remarkable rate. How has the response been so far on CREATIVE MARRIAGE and A GUIDE FOR RATIONAL LIVING? They should have a large audience and I wish you success with them.

Before closing I should note that little David is now two years and is quite proficient at keeping Harriet and I occupied fully. Right now he's pulling several of my books off the bookcase, so I best close and try and teach him some respect for property, at least for my property.

Best Wishes,

ALBERT ELLIS, PH. D.
PARC VENDOME
333 WEST 56TH STREET
NEW YORK 19, N. Y.

June 28, 1961

Dr. Ira Reiss
Bard College
Annandale on Hudson, NY

Dear Ira:

I have nothing directly to do with the new journal of sexual research, since Hugo Beigel is now the editor in chief, and I have too many other SSSS duties to do editorial work on it. You had better check with Hugo to see if he has your paper on hand, and tell him that you want DIANA changed to ATHENA. Otherwise, you might have to re-write Greek history!

I don't subscribe to the ANNALS any longer; so if you get reprints of your paper on the sexual codes of teenagers, by all means send me a copy. CREATIVE MARRIAGE is not selling phenomenally yet; but I think that the GUIDE is doing all right.

I am sorry that I have not been able to get together with you and Harriet recently; but it is quite possible that we will see more of each other when you are living much further away! My best regards to her and to dear little obnoxious David.

Most cordially,

ALBERT ELLIS, Ph. D.
PARC VENDOME
333 WEST 56TH STREET
NEW YORK 19, N. Y.

February 6, 1962

Dr. Ira Reiss
Dept. of Sociology & Anthropology
Iowa State University
Iowa City, Ia.

Dear Ira:

Many thanks for the reprint, which I shall get around to reading in short order.
This year—in fact, tomorrow—I am making the rounds of the Southwest—Dallas, Houston, Kansas City, and Fayetteville, and speaking at universities there. So I shan't get anywhere near Iowa City. But I do hope to get back there one of these days, since I found the psychological atmosphere there quite congenial.

On Friday, March 23, there will be a SSSS meeting at the Barbizon Plaza; but this will probably be after you have already departed from New York. I don't normally get to the Child Study Assoc. Meetings myself, since I find most of the people there insufferably stuffy; but if you and Harriet have any time when you are in New York, by all means give me a call and stop by for a drink or dinner. On weekdays, I am still terribly tied up until about 11 PM; but on weekends things are a bit less hectic.

Cordial regards,

Comments by Albert Ellis on His February 6, 1962 Letter

"I am still terribly tied up until 11 PM." Yes, at that time I saw my psychotherapy—and sex therapy—clients from 9:30 or 10:00 AM until 11 PM and on Saturdays until usually 6 PM. So I was goddamned busy! But I still did considerable reading and writing almost every week—especially when I took out-of-town planes and did some amount of writing at airports. As I said in an American Psychological Association newsletter around 1951, when asked how I wrote more than any other psychologist: "I have three advantages over other psychologists. First, I have an electric typewriter and I am a fast typist. Second, I—right now—have no helpmate, no girlfriend. Third, I am a non perfectionist and send practically everything I write hot off the typewriter, to publishers, with few corrections, except when editors demand them. For in the same time I could laboriously correct one of my pieces, I could normally write an additional one. I'm allergic to perfectionism!"

ALBERT ELLIS, PH. D.
PARC VENDOME
333 WEST 56TH STREET
NEW YORK 19, N. Y.

June 4, 1962

Dr. Ira Reiss
Dept. of Sociology & Anthropology
University of Iowa
Iowa City, Ia.

Dear Ira:

Many thanks for the preview of your chapter on your current NIMH research. I enjoyed reading it, and got some very useful information from it. Your scale on sex standards is very worth working out, and I am sure, as you note in your paper, that others will use it for their own researches. Please keep me posted on later research results.

It is good to know that you and Harriet enjoy Iowa. I like the intellectual climate of the midwest myself, and am about to take off this Wednesday for a trip to Minneapolis and the State Hospital at Hastings, Minn., to give a talk and a two-day workshop in rational-emotive psychotherapy. I find psychologists in the midwest more congenial to my approach than in most other sections of the country. I think they're saner and more scientific!

I have also been jaunting around other sections of the country, including the Southwest, Boston, Atlantic City, and Toronto, giving workshops and talks. So this has been a very busy year for me; and the converts to the cause are mounting.

I am sure that you are looking forward eagerly to your Brat the Second. I hope that all goes splendidly and that Harriet breaks all Special Delivery records!

Cordial regards,

Al

Oh, yes: my new books are coming off the press, and should be wafting your way in a couple of weeks or so.

Comments by Albert Ellis on His June 4, 1962 Letter

Because New York City was heavily psychoanalytic at that time—and still unfortunately is!—I managed, in spite of this, to do well in my New York practice, workshops, and lectures. But I found "Midwest empiricism" more to my liking, and was received better outside than inside New York. Today, I and other cognitive-behaviorists have appreciably helped to loosen up New York. Great!

October 27, 1965

Dear Al,

It was good seeing you in Canada, If you recall, we spoke of the likelihood of achieving orgasm in marriage and how this was related to premarital experience. In your talk at NCFR you said it was twice as likely to experience orgasm if you had premarital experience. I questioned that and I am writing to present the data sources. You referred to Kinsey and so am I: Human Female volume p. 406; shows that the failure in orgasm during the first year of marriage for experienced females who did not achieve orgasm in their premarital experience is not lower (in fact it's slightly higher) than the marital orgasm failure rate for virginal females who never achieved orgasm from any source. In short, the correlation that Kinsey found was *not* between having premarital intercourse and reaching orgasm in marriage but was between having premarital orgasm (from any source) and having marital orgasm. There is practically no relation between the simple act of having intercourse and the achievement of orgasm in marriage. A table summarizing this can be found on p. 185 of my book. If you have other data from Kinsey, or feel I am missing something here, please do let me know. I look forward to hearing from you.

Cordially,

Ira L. Reiss

THE INSTITUTE FOR RATIONAL LIVING, INC.
45 EAST 65TH STREET
NEW YORK N. Y. 10021

1 November 1965

Dr. Ira Reiss
Dept. of Sociology & Anthropology
University of Iowa
Iowa City, Iowa 52240

Dear Ira:

Many thanks for the information in your letter of October 27th. I can see that you are right about what you said and that I was not necessarily reading Kinsey but relying on my memory which is short and which apparently has neglected some of the essentials of the data that you have more accurately gone over. I shall keep in mind for the future, the points that you make in your letter and be more accurate in my quoting of some of the Kinsey material.

It was good seeing you again in Toronto, I had a large public meeting the next day after I saw you and on Monday morning I was able to have two full stories on my talks in Toronto in the Toronto papers. Since then some minister has taken up the cudgels against my coming out against pre-marital chastity and it looks like the fat is still in the fire in Toronto. Not that I object to this kind of a discussion in the public press—in fact, as you probably know, I welcome it, and on the whole I am very glad that I was able to stir things up there.

Cordial regards,

Albert Ellis, Ph.D.
Executive Director

Comments by Albert Ellis on His November 1, 1965 Letter

Yes, because I used Rational Emotive Behavior Therapy on myself, I was not taken aback by heavy criticism of me and my work in the area of sexual liberalism. I did not feel insulted by it or take it personally. I certainly never put myself down because of it. As I say in this letter to Ira, I was happy about this minister's criticism, since it brought my views to wide public attention and helped to liberalize some of the readers of the Toronto newspapers who didn't agree with the minister when he heavily objected to my liberal sex statements.

INSTITUTE
FOR
RATIONAL Living
INCORPORATED

October 4, 1967

Dr. Ira Reiss
Dept. of Sociology
Univ. of Iowa
Iowa City, Iowa

Dear Ira:

I was very pleased to receive a copy of your new book, "The Social Context of Pre-Marital Sexual Permissiveness." I think that you have <u>done an excellent research job</u> in this area, and certainly <u>one of a pioneering nature</u>. Some of the material that you point out on both general attitudes and sexual attitudes is <u>exceptionally valuable</u> and I am quite sure that your work <u>will spark a host of other researchers</u> to investigate some of the <u>most interesting hypothesis</u> that you have raised.

I could hardly find anything to quibble with in your presentation. However, I did wonder about the next to the last sentence on page 180 of the book. In this sentence, you say, "Indeed, if the question of how society takes large groups of people who are relatively equal in their innate sexual drives and teaches them to be so different in their attitude and behavior, can be answered, then we will have come along way to understand human society in general." I would wonder, first of all, whether the innate sexual drives, of our males and females, really are that relatively equal. I would suspect that the young male drive, at least is quite stronger, than the young female's, although later in life, this may not be quite so true. Secondly, although you consider the question in other regards in the book, at this point, you do not consider the possibility that the female of the species may have an innate drive toward conservatism that is rather greater than that of the male. If this were true, and the female were really more cautious about taking risks than the male, then it would appear that society's teachings are not as important as they appear on the surface, but that society very easily indoctrinates the female, while it has much more difficulty indoctrinating the male. Anyway, you might give some thought to this hypothesis.

Everything is going very busily with us at the Institute, and we keep adding to our facilities and our activities all the while. I hope that all goes quite well as usual, with you and your family.

Cordial regards,

Albert Ellis, Ph.D.
Executive Director

Comments by Albert Ellis on His October 4, 1967 Letter

I note that in this letter, written in 1967, I believed that human sex—and other—urges stem from biological *and* social learning influences. I definitely believe this today, and have written that human tendencies toward emotional and behavioral disturbance are both learned *and* acquired. Recent findings by evolutionary and psychoneurological psychologists have tended to agree with my views in this respect.

References

Advances in Sex Research: A Publication of the Society for the Scientific Study of Sex. 1963. New York: Hoeber Medical Division, Harper and Row.

Ellis, Albert. 1962. *Reason and Emotion in Psychotherapy.* Secaucus, N.J.: Citadel.

Ellis, Albert, and Robert A. Harper. 1961. *Creative Marriage.* New York: Lyle Stuart.

———. 1961. *A Guide to Rational Living.* North Hollywood, Calif.: Wilshire.

Reiss, Ira L. 1960. *Premarital Sexual Standards in America.* Glencoe, Ill.: Free Press.

———. 1967. *The Social Context of Premarital Sexual Permissiveness.* New York: Holt, Rinehart and Winston.

Overall Changes in My Views 5

ALBERT ELLIS

As shown in my comments on my first letters to Ira, we often differ in terminology than in the views that we stated. To this day, I still hold that both sexual and affectionate desires have strong biological elements but that these are intrinsically and "holistically" integrated with important social learning. Because sexual urges in our culture—and especially in the 1950s—were denigrated and censored, I strongly encouraged their verbal-physical expression more than Ira at first did. But I never deified sex desire and always placed it, in my writings and my Rational Emotive Behavior Therapy, with a sensible and responsible general values system or philosophy.

I have personally always been "addicted" to romantic love, and have been "in love" with over two dozen girls and women during my life. But, from my own experiences and from my wide reading on love, I have steadily gone along with Stendhal's view in his nonfictional book *On Love* (1947) that although what I would call "loving kindness" may last for many years, romantic love almost always wanes after a few years of major living-together-arrangements (LTAs). Therefore, lovers had better not take it too seriously as a prime reason for their marrying. Discussing this matter with my clients has saved many romantic souls from disastrous marriages!

As I note in my comments on my letter to Ira of February 22, 1957, I still strongly uphold my own and other people's individualism, but I place it on an equal plane with what Alfred Adler called "social interest." As long as we humans choose to live in families and communities with other people, human individuality and social responsibility—both/and, not either/or—are the sensible ways to go. As Patrick Henry aptly said, "Either we hang together or we hang separately!"

My point, in my March 9, 1957 letter, that people "must never, I insist, blame themselves for their past behavior," is a crucial aspect of my personal philosophy and of REBT. This point is central to Carl Rogers's Person-centered Therapy as expressed in his book *Becoming a Person* (1961) and also to Rational Emotive Behavior Therapy. It was emphasized in my earliest book on REBT, *How to Live with a Neurotic* (1957). Although I am a very active-directive therapist and Rogers was much more passive, we both agreed that even when people act "stupidly" and "badly," they never have to condemn their self or being but merely had better negatively assess their behaviors. That is an essential value system of REBT and it is consistently expressed in my letters to Ira. Fortunately, though he at times definitely disagreed with my sexuoamative values, he consistently showed that he never damned other people for their views and behaviors.

As I note in my comments on my letter to Ira of March 23, 1957, today I am somewhat closer to his view of people conforming to sexual and other social standards. After using REBT with thousands of individual and group therapy clients, I realize that just about all conventions have their advantages and their disadvantages, both of which better be considered before people chose conventional or unconventional pathways.

Again, in my comments on my letter of April 2, 1957, I somewhat revise down my extreme sexual liberalism of that year. I put it in the context of people being more responsible and less impulsive (not to mention, compulsive) in their general individual and social living.

The Declaration of Interdependence that I sent to Ira with my letter of April 16, 1957 and that Rhoda and I enthusiastically signed, helped us temporarily—but hardly permanently. Two and a half years later we were happily divorced! But more than forty years after our divorce, we're still good friends. Our "in-lovedness," as I noted before, was not made to last. But our "loving-kindness" continues!

As I noted in my comments to my letter to Ira and Harriet of April 28, 1957, *The Folklore of Sex* (1951) was killed by Doubleday. My four next books, *Sex Life of the American Woman and the Kinsey Report* (1953), *Sex, Society and the Individual* (1953), *The American Sexual Tragedy* (1954), and *The Psychology of Sex Offenders* (1956), were also not promoted or were deliberately censored by their publishers and didn't sell very well. But I managed to use my developing philosophy of Rational Emotive Behavior therapy to stop from killing the publishers or myself!

My comments on my letter of June 2, 1957 show that I was somewhat dogmatic about religion having no place in ethics. I now see that religion often is quite ethical, and that religions that include a kindly and forgiving, instead of a damning and vindictive god, are particularly ethical and have, in spite of their dogmatic dictates, contributed significantly to social morality.

In my comments on my letter to Ira and Harriet of August 21, 1957 I show that I am happy that REBT, when it was in its infancy, distinctly espoused the therapeutic philosophy of unconditional self-acceptance (USA) and unconditional other-acceptance (UOA). It still strongly endorses these aspects of mental health today!

In my comments on my letter of September 7, 1957 I again admit that at that time I was too rigid about claiming that all devout religion is mentally harmful. Today, I realize that some forms of religion can aid the emotional health of devout believers. Details of my present and more favorable views on religion can be found in my recent book with Steven Lars Nielsen and W. Brad Johnson, *Counseling and Psychotherapy with Religious Persons: A Rational Emotive Behavior Therapy Approach* (2001).

In my comments on my letter of November 10, 1957 I agree with the existentialist position that people's having a central meaning in life is favorable to their mental health. In fact, Robert A. Harper and I enthusiastically endorsed this position in *A Guide to Rational Living* (1961), and we particularly encouraged our readers to acquire a long-lasting vital absorbing interest. As I point out, however, and still tend to think today, their vital absorbing interest had better be personally and preferentially chosen and not merely selected because of social conventions.

As I note in my comments on my letter of November 24, 1957, I tone down my somewhat rabid individualism of that day and realize that my nonconformism suits me very nicely, but may well not be best for other people. Today, I am frankly still biased in favor of my own individualism, but I do my best to exert it within a social interest framework. In this respect, I am somewhat like Epicurus—who strongly espoused personal pleasure, but within a disciplined framework. I definitely agree. Both Epicurus and I follow Alfred Korzybski's principle of viewing human behavior in a both/and, instead of an either/or, outlook.

As indicated in my comments on my letter to Ira and Harriet of August 6, 1958, I still give myself credit for my persistent efforts to organize the Society for the Scientific Study of Sex—now called the Society for the Scientific Study of Sexuality. Since its inception, the SSSS has done more to promote sex research than any other group—and has always maintained an open-minded, flexible attitude toward the many varieties and aspects of human sexuality that are relevant. Both Ira and I for many years have served the SSSS with distinction.

As I noted in my comments on my letter of January 4, 1959, I thought that the *Encyclopedia of Sexual Behavior* (1961) was one of my best publications. It promoted pioneeringly liberal attitudes toward sex, love, and marriage—and I am still proud of compiling and contributing to it.

In my comments on my letter of August 14, 1960 I tend to agree with Ira that "rational" has several possible meanings and is never absolutistic or rigid. In REBT, "rational behavior" means that which helps you and your social group achieve what you consider useful results. But it is a somewhat slippery term and has no fixed and universal meaning.

In my comments on my letter of October 7, 1960 I revise my notion that American sex standards are "very arbitrary" and are "often unrealistic." I acknowledge that such standards, although they "arbitrarily" restrict people from performing acts that would be personally beneficial, make some sense in terms of general human functioning and survival.

In sum, these letters to Ira cover eleven years between 1956 and 1967 when the sex revolution of the late 1960s was just getting started and later was in full swing. They show how both Ira and I were in the vanguard of that revolution and in many ways did our best to get it going. While we were doing so, we significantly differed in some of our major views on sex, affectionality, rationality, religion, and other social issues. But we consistently heard each other out—and both of us seemed to benefit considerably by doing so.

Resulting from our interchange of these letters as well as from our researches and experiences since 1967, Ira and I have continued growing, as some of our comments on our letters duly attest. Although I always prided myself on my open-mindedness to the nature of humans and the psychological and social manifestations, I think that I have become increasingly so over the last half century. The relatively few rigidities I had in the 1950s and 1960s have become more flexible. Let us hope that I keep chipping away at them further in the years still to come!

References

Ellis, Albert. 1951. *The Folklore of Sex*. New York: Charles Boni.

———. 1953. *Sex Life of the American Woman and the Kinsey Report*. New York: Greenberg.

———. 1953. *Sex, Society and the Individual*. Bombay: International Journal of Sexology.

———. 1954. *The American Sexual Tragedy*. New York: Twayne.

———. 1956. *The Psychology of Sex Offenders*. Springfield, Ill.: Thomas.

———. 1957. *How to Live with a Neurotic: At Home and at Work*. North Hollywood, Calif.: Melvin Powers.

Ellis, Albert, and Albert Abarbanel, eds. 1961. *Encyclopedia of Sexual Behavior*. 2 vols. New York: Hawthorn.

Ellis, Albert, and Robert A. Harper. 1961. *A Guide to Rational Living*. North Hollywood, Calif.: Wilshire.

Nielsen, Stephen Lars, W. Brad Johnson, and Albert Ellis. 2001. *Counseling and Psychotherapy with Religious Persons: A Rational Emotive Behavior Therapy Approach*. Mahwah, N.J.: Erlbaum.

Rogers, Carl. 1961. *Becoming a Person*. Boston: Houghton-Mifflin.

Stendhal. 1947. *On Love*. New York: Liveright.

Taking Stock of My Past and Present Views   6

IRA L. REISS

I WILL TRY TO SUMMARIZE and integrate changes and similarities in my thinking today as compared to my views expressed in my letters. I will start by going over some key issues between Al and I that came out in our letters and then present some overall comments on my work over the years.

In chapter 1, we debated at some length in our letters the worth of casual sex. I gave less value to and saw more risk in casual sex than Al did. My views today are more acceptant of casual sex. As we begin the twenty-first century, our society clearly has changed and is much more acceptant of individual sexual choices than it was in the beginning of the second half of the twentieth century. So the emotional and guilt consequences of all kinds of sex will generally be less than they were then. Furthermore, our young people are more knowledgeable about sexuality. Even our teenagers are catching up with teens in other Western cultures by improving their use of condoms to protect themselves against disease as well as pregnancy.

If someone wants to pursue casual sex as his or her primary or only goal, I would not now see that as too narrow. I still personally believe in the higher value of a "person-centered" sexual relationship, but I don't feel that that perspective has to be endorsed by everyone. In this sense, I have broadened my pluralistic stance and more fully accepted casual sex as long as it is freely chosen and nonexploitative. I think Al was already in this more pluralistic stance when we wrote our letters. Surely, I never condemned casual sex, but I did not give it more than a minor value.

In my ethical thinking today, I am more focused on a pluralistic and minimalist ethical stance that rejects only forced sexuality, such as rape, and exploitative sexual relationships, such as adult–child sex, but accepts all other sexuality that is free from force and exploitation. I have published two books explaining how such sexual pluralism is needed if we are to learn how to minimize our society's many sexual problems. The most recent one is *Solving America's Sexual Crises* (1997). I feel that people in general are moving toward such a more pluralistic sexual position and that is why our society is becoming better at containing sexual problems. I believe the 1998–1999 public reaction to the "Bill and Monica" scandal indicated this pluralistic trend. The public didn't like Clinton's behavior but they didn't allow that to change their positive opinion of the work he was doing as president. This was a significant change from the 1987 Gary Hart scandal that ended his presidential bid.

Al and I had differences in the importance we placed on biology as a reason for more fully accepting sexuality. I accepted a rather wide variety of sex but not because of any biological "drive" to satisfaction that Al discussed in his letters. Biology played only a minor role in my theorizing. I still see biology as setting very broad limits and as being malleable to a considerable extent. This biological flexibility is clearly evidenced in the many changes in American premarital sexual attitudes and behaviors in just the past several decades. In recent years, the evidence of specific biological inputs has increased and so I do now accept more of a role for biology in sexual relationships. I still am predominantly interested in sociological explanations and I do believe in the primacy of learned behavior and attitudes in our sexual lives. In my mind, society supplies the software that runs the biological hardware and it can make that same hardware do a very wide variety of things. In sum, I see an interactive relationship between biology and society and perceive it as quite flexible and variable. Al's current comments on his letters in this book indicate that he has lessened his stress on biology, so our positions today may well be a great deal closer than they were forty-five years ago.

In chapter 2, religion was an area where we differed considerably in our letters. Al was less acceptant of any organized religion. I surely would not then or now try to persuade anyone to join any organized religion and we were in agreement on that. Also, I, too, would reject a religion that stressed what Al most disliked: "self-blame and other blame." I have always rejected religions that claim to be the only "true" religion. But I was then and still am convinced that there are religions like Reform Judaism and Unitarianism and increasingly many others that allow for identity with a tradition but also allow for the fresh air of reason and evidence to enter into the temple or church. So I would push for reforming religion and accepting people's desire for an organized place in which to develop their thoughts and feelings about the meaning of life. To me, organized religion can be such a place, even if it often fails to achieve this goal. Here, too, in his current comments on specific letters Al indicates that he has significantly softened his negative views about organized religion and so our views about religion have also converged.

I would add that I still believe in the existence of some sort of God. But my God does not wear a yarmulke or a cross or the insignia of any other specific religion. God to me is not contained in any one religion. I see God as a kind of "first cause" of the universe. I still like the deistic view of a God that set the world and humanity in motion and gave us the potential for helping ourselves and each other to make life a rewarding, meaningful experience. But it is predominantly through our efforts, and not direct divine intervention, that we can make progress to a better society. On more specific religious beliefs like the belief in an afterlife, I generally take an agnostic position—I just don't know. Overall, I still accept a progressive, nondogmatic form of organized religion as a good and necessary part of human society. Religion of this sort in our society supports the ethics of civil society and can challenge antidemocratic movements and question the political and economic powers that be. Nevertheless, I see those religious groups that present a narrow, nonpluralistic view of morality as counterproductive to the good of society but, of course, they have a right to express their viewpoints.

A point of contention in our letters in chapter 3 concerned just how dependent the individual was on society. I strongly support individual traits of independence, irreverence, and speaking one's mind. I joined the American Civil Liberties Union in 1948 while I was in college and I have been a member ever since. I most strongly support our first amendment rights to express our views, no matter how distasteful to others they may be. In my own life, honesty and forthrightness has most often trumped propriety, sometimes to the chagrin of my family and friends. Independent individualistic values are a central part of my own personality. But I also believe that the attitudes of the significant others in our lives are very important to us. Of course, we can decide to go in our own direction because that is of great personal importance and we can expect that others, if they are pluralists, will not begrudge us this freedom. Surely, we each should give each other broad room to make our own choices and to express the kind of person that we are. But just as surely we also have to attend to some degree to the needs of our close friends and family in our decisions. There is a balance here in which each of us must seek our comfort zone. After reading Al's current letter comments updating his perspective, I believe we are today closer on this area of individualism and dependency on others.

Chapter 4 illustrates one area where I have accepted a wider range of sexual standards than Al did. This relates to judgments about abstinent people. Al felt that abstinent practitioners must be "anxious or hostile" in order to maintain their abstinence. To be honest, I always put abstinence at the bottom of my list of acceptable sexual choices. In this sense, I agreed with the twelfth-century romantic love writer Andreas Cappelanus, who said that the woman who is abstinent in life will go to the least happy place in the afterlife (*The Art of Courtly Love* [1959]). But

I still accepted the fact that some people might well find abstinence most congenial—at least for a portion of their life. Today, I still feel the same way. Abstinence that does not involve force or exploitation is within the moral ballpark of my sexual pluralistic ethic. Nevertheless, I believe it is very difficult to maintain in today's society. I would not promote it but I also would not see it as inevitably harmful.

I should mention that over the years things have changed some in my areas of interest within sociology. At the time of the letters in this book, I was working on my first book (1960) and then doing the research for my second book (1967). Both of these books were on premarital sexual standards in America. Surely, premarital sex is still an important area of my interest. But my interests have expanded greatly over the years as my autobiography in appendix B spells out. I have written about extramarital sexuality, cross-cultural comparisons of all forms of sexuality, major sexual problems like rape, HIV/AIDS, child sexual abuse, and teenage pregnancy, and also on problems in sexuality research and theory. In addition, I wrote a textbook on the family wherein I highlighted the connections of the family to various sexual areas of our lives (*Family Systems in America* [1971]). I believe the success of the four editions of this text helped encourage more sexuality coverage in the family textbooks that followed mine. So my work in the family was also tied in with my broad primary interest in studying and explaining human sexuality from a societal point of view.

But even though my substantive interest has expanded, one thing has been constant: My major enjoyment and goal in my professional work was and is to build sociological explanations—to develop theories that could explain how society impacts sexuality and try to formulate predictions and ways of changing society based on these theories. I have tested my theories using national survey samples as well as by other more qualitative methods. A number of my predictions have held up rather well. For example, in my 1960 book I predicted the sexual revolution of the 1960s and more recently predicted many of the changes in sexual behavior and attitudes that took place in the 1990s (1960, 239–241; 1997, 14). Building explanations remains my first love and this fascinating search for understanding and scientific explanation is probably a key basis for why I like to discuss controversial areas of sexuality. I feel I have helped develop ways of dealing with such controversial questions—at least partially and temporarily.

I must add that although I do greatly value the scientific approach, I reject the narrow positivistic view of science that was so popular in the first half of the twentieth century. Rather, I see science as the best way to become aware of our own potential biases and those of others. It affords a way of getting beyond just asserting what you think or what you like and instead carefully examining the evidence and reasoning involved. I believe that the letters between Al and I in this book indicate exactly how this process of mutual questioning and challenging builds a stronger and better scientific explanation. Although I do not hide my personal views, I have tried very hard to keep them from biasing my analysis of data and my development of theory. My personal values do lead me to study specific sexual problem areas and they do guide me in selecting my preferred solutions to those problem areas. But science emphasizes the need to test and examine your ideas so that your personal values, while influential, are less likely to bias your understanding. The scientific attempt to examine controversial issues in as fair a way as possible has always appealed to me. I mention this regard for a broad conception of science because I want it to be clear that despite the very personal comments made in this book and in our letters, I do strive for fairness and balance in my professional views about how society shapes our sexuality. I feel sure that Al does precisely the same.

I would like to end this chapter by noting that reading Al's writing about sexuality issues was a major source of satisfaction and support. The world, especially during the 1950s, was not generally too friendly to sexual liberals. College faculties are almost always among the most liberal people, but in those early years even they would mostly tolerate but not endorse sexual liberals. But in Al I found someone who was very much like me—only more so. We both saw a great deal wrong with our society's way of handling sexuality. We may have differed on specifics, but we fully agreed that things had to change and to change a great deal before our sexuality could be celebrated rather than denigrated. It has been a pleasure and a comfort to have known Al so early in my career and to see his therapeutic approach become so widely appreciated.

Al took a more individualistic and therapeutic stance and I took a more societal and research approach. But our two paths were most often parallel to each other and mutually reinforcing. I believe our choice of fields was an indication of our deepest personal interests and that the professional work accommodated those basic personality traits. But underneath our specializations, we shared a great deal of our fundamental values supporting independence, freedom of choice, intellectual curiosity, and valuing love and pleasure. I am sure that if we wanted to, Al and I could enter into lengthy discussions about differences even in our current views, but we now know that in our most basic values, we are cut from the same ethical mold.

Al and I are people who want to build a better society and make the world friendlier and kinder to those in it. We also seek to examine our ideas carefully through reason and scientific research so that we don't substitute a personal bias for a societal bias. You can see this willingness to change in our letters and also in our comments about what we currently believe. I would say that in present-day American and Western society the sexual pluralism that Al and I endorsed many decades ago is increasingly winning the day. This broader social and personal acceptance of the positive aspects of sexuality and of the right of people to chose is a major good for our society. I like to think that Al and I had something to do with this outcome.

References

Cappelanus, Andreas. 1959. *The Art of Courtly Love.* New York: Ungar.
Reiss, Ira. 1960. *Premarital Sexual Standards in America.* Glencoe, Ill.: Free Press.
———. 1967. *The Social Context of Premarital Sexual Permissiveness.* New York: Holt, Rinehart and Winston.
———. 1971. *Family Systems in America.* New York: Holt, Rinehart and Winston. (2nd ed. 1976; 3rd ed. 1980.)
———. 1997. *Solving America's Sexual Crises.* Amherst, N.Y.: Prometheus.
Reiss, Ira L., and Gary R. Lee. 1988. *Family Systems in America.* 4th ed. New York: Holt, Rinehart and Winston.

Autobiographical Appendices

How I Became Interested in Sexology and Sex Therapy

ALBERT ELLIS

Albert Ellis was one of the founders of the Society for the Scientific Study of Sex and its first president. He is also the founder of the Institute for Rational Emotive therapy in New York City and is the best-known practitioner of Rational-Emotive Therapy (RET), a form of cognitive behavior therapy.

MY KEEN INTEREST IN SEX BEGAN, at the very latest, at the age of five, when I was caught by my parents trying to pour some milk through a funnel into the vagina of Mary J., a blonde-haired and blue-eyed five-year-old bombshell with whom I was madly in love. I don't remember that I considered this act sexual. But both our sets of parents certainly did! So Mary, my beloved and my friend, was abruptly whisked out of my life.

I continued sexual—or at least nude—exploration at the age of seven when I spent ten months with nephritis in New York's Presbyterian Hospital and used a flashlight to reveal the nude bodies of the other children in my ward—and to let them see how scrumptiously I was hung. Still curiosity—no real sex.

However, I soon graduated, and just about got my first degree in sexual pleasure, when I discovered that if I pressed my genitals against the rails of my hospital crib, I could get a semi-orgasm—a real sexy thrill. So I made myself slightly addicted to crib-rail sex. Later on, when I went back to school, I found that climbing ropes in gym could also lead to genital joy. So, though I generally hated gym, I became quite a rope climber!

My first feeling of shame about sex was at the age of twelve, when I found that I had an almost constant erection, and was afraid that other people—especially the girls in my class whom I obsessively loved—would notice it and despise me. Oddly enough, however, I was still so ignorant about sex that I didn't start to control my perpetual erections by masturbating until I was fifteen. Then I really went to town!

I masturbated twice daily and wasn't guilty about that until I figured that maybe I was too uncontrolled. So I rushed to the public library—to which I was also addicted—and found a few books that said that masturbation was okay, even when frequent. Pretty good books for the 1920s!

From the age of sixteen onward (in 1929), I read many books by Freud and his followers, but I could see that Freud was especially obsessed with the sexual "origins" of disturbance, especially with the ubiquitousness of the Oedipus complex. I could also see that he was an overgeneralizer and a dogmatist, and therefore a poor scientist. But I was helped by psychoanalytic details about sex to loosen up and to consider practically all forms of noncoercive sex permissible. In fact, at the age of fifteen, I had my first and only homosexual episode—with my thirteen-year-old brother no less! I saw that I could easily come to orgasm in that way, too, and didn't feel one bit guilty. Realizing, however, that to seek out gay sex would probably get me into trouble—especially in *those* days!—I thereafter continued to obsess about and to try to have sex only with women. That was a little easier and safer.

Sidelight: When Alfred Kinsey, about fifteen years later, interviewed me for his first report and discovered that I was highly heterosexual, he seemed delighted to find that I also had had an early homosexual encounter. I think that he was determined to discover that a large percentage of males had at least one homosexual episode in their lives. So fortunately I fulfilled this quota!

My accompanying obsession to sex was writing. At the age of twelve, I decided I would mainly be a writer; and between my eighteenth and twenty-eighth years I wrote no fewer than twenty book-length manuscripts—novels, plays, poems, and nonfiction works. Many of them, especially my novels, were very sexy. In fact, probably too sexy to be published—for I described sex play as few published writers other than James Joyce had previously done.

No dice. I got a number of fine rejection letters, and some near-publications, but no real hits. So at the age of twenty-six I decided that nonfiction writing on sex, love, and marriage would be my best bet. Why? Because I was quite interested in these related fields; and presumably because those areas sold well.

As usual, I went to the New York Public Library, and to two special private libraries I was a member of. Between all three, I could borrow ten books every day. So I did—well, at least five days a week. On Saturdays and Sundays I went to the main reading room at the 42nd Street Public Library, and devoured from thirty to fifty books each day.

Thirty to fifty? Yes, indeed. I am a fast reader. Without having taken any Evelyn Woods courses, I skim magnificently. Especially when all the books are on sex, love, and marriage, and most of them nauseatingly repetitious.

For two years I read hundreds of books and thousands of articles and began writing my "masterpiece": a thousand-page, single-spaced tome titled *The Case for Promiscuity*. Did I get any publication offers on it? Not one. Some great feedback from editors, but the unanimous view was that it was just too liberal for the early 1940s. Years later, I published the first volume of it (Ellis, 1965).

My notable library endeavors, however, catapulted me into becoming a sexologist and a sex therapist. My friends and relatives, hearing that I was studying sex, love, and marriage so intently, started to ask me how they could deal with their personal problems in these important areas. To my surprise, I was able to give them some good answers. In one or a few (unpaid) sessions of avid and often highly intimate conversation, I dispelled their ignorance, gave them realistic suggestions, relieved their anxiety, and helped a number of them to lead happier sex, love, and marital lives.

My subjects and I both benefited. I discovered much additional sex-love information. And I greatly *enjoyed* doing this kind of counseling, so much so that at twenty-seven I funded the Love and Marriage Problems (LAMP) Institute, devoted to helping and to doing research in sex, love, and marriage.

Great! But I had no status in the field, and my lawyer for my first divorce strongly advised me to get some. At that time, graduate schooling in sex therapy and even in marriage and family therapy did not exist. The closest thing to it was a Ph.D. degree in clinical psychology. So I applied for that.

Because my bachelors degree was in business administration—which I really took to make money and support myself as a writer—I had trouble getting into graduate school at NYU and Columbia. But I did so well in taking three trial courses during the summer term at Columbia that they broke down and let me enroll in the clinical psychology program at Teachers College for the fall of 1942.

A piece of cake! Within a year, by *not* letting Teachers College know that I was working full-time, I received my M.A. with honors and was soon matriculated for a Ph.D. degree

Then came, however, a hitch! When it was time to work on my doctoral dissertation, I had the good sense to avoid doing one on any overtly sexual subject. That would have been verboten. So I thought I skirted this ticklish issue by doing one on *The Love Emotions of College-Level Women*. Everything at first went well and I was about to write up my interesting data on this subject when—wham!—Teachers College, Columbia, most unusually forced me to hold a special seminar to decide whether my topic was kosher.

Whereupon all thirteen members of the clinical psychology department got together with the provost of the college. The provost? I didn't realize that we even had one, until my main advisor told me that we definitely did—for administrative and not academic purposes. He never sat in on a thesis seminar. But this time he had heard about my topic, was afraid that the Hearst newspapers would make a federal case of it, and called this special seminar to see if I was to be allowed to polish it off.

Nothing daunted, I presented my research to the assembled professors and the provost, they politely listened, and then voted twelve to two in my favor. They agreed that I had a great topic, that I was a fine person and a scholar, and that I was proceeding scientifically with my study. Marvelous! But, as my advisor, Professor Goodwin Watson, sadly told me, the two (anonymous) dissenters were going to adamantly oppose this and *any* dissertation on love that I could come up with. They would be sure to show up at my final thesis orals and knock me for a loop, no matter *what* I did.

Well, that ended that. Not being a whiner, and wanting my degree, I picked another topic that had nothing to do with love, sex, or marriage, and quickly polished off a safe, highly statistical dissertation, *A Comparison of the Use of Direct and Indirect Phrasing in Personality Questionnaires* (Ellis, 1947a), which was nicely accepted. My poor orphaned thesis on love? I turned it into seven pioneering articles on the subject which I published in several psychological and sociological journals. So I fixed Teachers College, Columbia (Ellis, 1947b, 1948, 1949a, 1949b, 1949c, 1950, 1953d).

Soon after this, I began writing on sex with a vengeance; by 1954 I had published no fewer than forty-six articles on sex, love, and marriage, in addition to two books (Ellis, 1951, 1954a) and two anthologies (Ellis, 1954c; Pillay and Ellis, 1953). I was also the American editor of the pioneering journal the *International Journal of Sexology*. I had a wide psychotherapy practice and specialized in seeing hundreds of clients with sex, love, and marital problems.

Although I was well known in the psychology profession as a sexologist in the 1950s, my public fame mushroomed in the early 1960s when several of my books appeared in paperback form and became bestsellers. Thus, millions of copies were sold—and more millions borrowed from libraries—of my books *Sex without Guilt* (Ellis, 1958), *The Art and Science of Love* (Ellis, 1960a), *The Intelligent Woman's Guide to Man-Hunting* (Ellis, 1963a), *Sex and the Single Man* (Ellis, 1963b), and *Nymphomania: A Study of the Oversexed Woman* (Ellis and Sagarin, 1964). Although at this time my name was not exactly, like Kinsey's, a household word, his books were widely bought and not read while mine were avidly read as well as bought. I still, thirty years later, meet many individuals who enthusiastically tell me that they were first catapulted into enjoyable and guiltless sex by reading my books. I think that I can say without immodesty that the sexual revolution of that decade was largely sparked by the writings of Kinsey and Ellis.

All this notoriety at first did me little good professionally. To be sure, I received many referrals for sex therapy from psychologists and sexologists whom I had never met and who knew about me only from my publications. But I also received much opposition from professionals who were against popular writing by psychologists and who therefore hated my guts.

Professional journals and other publications have actually censored some of my sex writings. Thus, the *International Journal of Sexology* refused to publish my article "New Light on Masturbation" (Ellis, 1956a) because it would offend certain members of sectarian groups if Dr. Pillay, the editor, published it in India. The *Journal of Social Therapy* also refused to publish my article on masturbation among prisoners. The editors of the book *Sexual Behavior in American Society* refused to republish a pro-Kinsey article of mine that they had already published in the journal *Social Problems*.

For many years national popular magazines, such as *Redbook* and the *Ladies Home Journal*, asked me to do articles on sex; but when I sent them outlines or actual articles, they found them "too unrealistic," "too bold for our readers," or "too controversial." *Esquire* accepted one of my articles, "A Case for Polygamy" (Ellis, 1960b), and paid me well for it. But one of their chief editors later found it "too strong" and refused to publish it.

In 1958, after I published the first edition of *Sex without Guilt*, I appeared on many radio and TV shows, but also ran into considerable censorship. Although many of my Long John Nebel radio shows were very popular and were recorded and often rebroadcast, the show that I did on *Sex without Guilt*, in which I specifically mentioned masturbation and fornication as desirable acts, was forbidden to be rebroadcast by the management of radio station WOR.

I have had many other brushes with censorship and with being first invited, and then uninvited, to appear on popular radio and TV shows. Twice the Federal Communications Commission took the programs in which I was appearing temporarily off the air because of my espousal of premarital sex relations. On David Susskind's "Open End" TV show I appeared with Max Lerner, Hugh Hefner, Ralph Ginzburg, Maxine Davis, and Reverend Arthur Kinsolving for a two-hour presentation, "The Sex Revolution." But when Susskind asked me on the air what I would do if I had a teenage daughter who insisted on having premarital sex and I replied, "I would fit her up with a diaphragm or birth control pills and tell her to have fun," the tape that we made was banned from the air and never played anywhere. Ironically enough, Max Lerner, on this same show, had previously remarked that the mere fact that we were doing this program that night showed how liberal TV was becoming in its attitudes toward sex!

Several of my books have been banned here and abroad, especially *Sex without Guilt*, whose sale was prohibited in a county of southern California. The opposition to my work as a sexologist also spilled over significantly into my reputation as a theorist and practitioner of psychotherapy. In 1955 I started to do rational emotive behavior therapy (REBT, later RET), the pioneering form of cognitive-behavior therapy, and since that time I have spent by far most of my time as a therapist, writer, and workshop presenter dealing with general psychotherapy, not just with sex therapy. But my reputation as a sexologist followed me into the general field of therapy, too; I have often been savagely criticized for my "controversial"—meaning, largely sexual—positions.

Even friends and associates of mine who follow much of my therapeutic teachings have often tried to induce me to tone down my sexual writings and presentations. Thus, Rollo May, who used to send me his "difficult clients" for RET when he practiced in New York in the 1950s and 1960s, opposed my becoming president of the American Academy of Psychotherapists because I was "too controversial." And many other "respectable" psychologists have used my therapeutic teachings without giving me due credit, or even going out of their way to criticize me, largely because of my reputation as a liberal sexologist.

Some other prominent sexologists have also vigorously opposed my sexual liberalism and my widespread popular publications on sex. My good friend Hans Lehfeldt, who helped me found the Society for the Scientific Study of Sex in the 1950s, opposed my being nominated as its first president because he felt that, once again, I was "too controversial." Fortunately for me, the other members of our board of directors, including Harry Benjamin and Henry Guze, did not go along with Hans. So I was renominated and elected.

In spite of this kind of opposition, I continued to absorb myself in the field of sexology as well as psychotherapy, largely because I took my own counsel as a therapist and firmly taught myself that it is great to be loved and approved by other people, including members of one's own profession, but that it is far from *necessary*. This is one of the main tenets of RET. Without following it myself, I might well have given up being a sexologist and merely have stayed with the safer aspects of psychotherapy and counseling. On the other hand, clinical sexology definitely has its rewarding aspects which kept me interested in working within it despite the disadvantages that it also entailed. Let me mention some of the rewards that I have found in being absorbed in it for over half a century.

First, sexology is the science of sexuality. I greatly enjoy the *discovery* aspect of science. Thus, in 1943 I was doing a term paper for a clinical psychology class on the various "causes" of homosexual behavior, when I accidentally discovered that most hermaphrodites whose libidinous direction was known were heterosexual in spite of their physiological and hormonal anomalies. This led me to publish my first scientific paper, "The Sexual Psychology of Human Hermaphrodites" (Ellis, 1945), which appeared in *Psychosomatic Medicine* and created quite a stir. It showed that although the power or strength of the human sex drive is strongly influenced by biological factors, its direction is largely influenced by familial and cultural teachings.

This was a very exciting discovery; I enjoyed the article's being cited in a great many articles and books and its influencing the work of some other outstanding sexologists, such as Harry Benjamin and John Money. Almost any kind of discovery—including artistic, political, and economic findings—can be uplifting. This, of course, also goes for sexological discovery.

Second, I have always found that my absorption in clinical sexology has important practical aspects. From my earliest consultations with my friends and relatives to my later sessions with thousands of clients and my writings for therapists and for the public, I have apparently helped almost innumerable people with their sex, love, marriage, and general problems. This has been, and still is, most gratifying. The hundreds of voluntary endorsements that I have received from readers in every major country in the world have been especially satisfying. Conveying my knowledge to these people has often produced splendid results.

Third, I have naturally benefited myself from my sexological findings. Not only, as noted above, did I overcome my own guilt and shame about sex largely as a result of my reading—and not from conversations with mentors or therapists—but I also tackled some of my own problems, such as the fast ejaculation that I was plagued with early in my life, and significantly helped myself by my sex researches. In addition, I helped quite a number of my sex and love partners to decrease their anxiety and increase their pleasure. And that has been great!

If I had my life to live over again, would I still choose the field of sexology as one of my major pursuits? Definitely. I have always disagreed with Freud (1965) that sex problems, including incestuous thoughts and actions, are the major causes of general emotional problems, such as anxiety, depression, self-deprecation, and rage. One of the causes or contributions, yes, but hardly the only one. On the contrary, as I have said for several decades, general human disturbance is much more likely to lead to sex disturbance than vice versa. Actually, the two tend to be interactional.

Humans, as I keep preaching in RET, are both born and reared to be easily disturbable; and their emotional and behavioral disorders are both nonsexual and sexual. The solutions to their problems, moreover, are complex and involve a number of important cognitive, emotive, and behavioral insights and methods (Ellis, 1957, 1962, 1988, 1994, 1996). But sex is an exceptionally important aspect; and anything that we can do to minimize its disorders and enhance its fulfillment is meaningful and of great consequence.

The twentieth century has so far been the outstanding period of general psychotherapy and of sex therapy. In the latter area we have the pioneering findings and ideas of a number of unusual scholars and clinicians, including Havelock Ellis (1936), Alfred Kinsey and his associates (Kinsey, Pomeroy, and Martin, 1948; Kinsey, Pomeroy, Martin, and Gebhard, 1953), William Masters and Virginia

Johnson (1960, 1970), Joseph LoPiccolo (LoPiccolo, Stewart, and Watkins, 1972; LoPiccolo and LoPiccolo, 1978), and many others. I am delighted to have been one who was friendly with and who worked with these pioneering sexologists and to have made some significant contributions myself.

Shall I risk mentioning what I consider my main contributions to the field of sex, love, and marriage? Why not? Here are some of them:

1. I was probably the first prominent psychologist to unequivocally point out that masturbation is not only not harmful and shameful, but that it is also actually beneficial for most people (Ellis, 1951a, 1952a, 1954a, 1955a, 1956a, 1958).
2. I was one of the few psychologists in the 1940s and 1950s who told my clients and members of the public that mutual consenting premarital sex relations for adults, when engaged in with proper sexually transmitted disease and pregnancy precautions, are not necessarily bad or immoral and can enhance one's sexual and general life (Ellis, 1951a, 1953a, 1954a, 1955a, 1956b, 1958,1963a,1963b,1965, 1976; Ellis and Harper, 1961, 1975).
3. I vigorously opposed the idea that unconventional sex behavior is perverse or deviant and proposed that sexual "abnormality" is usually a myth (Ellis, 1951a, 1952b, 1952c, 1954a, 1954c, 1958, 1963a, 1963b).
4. I was a pioneering feminist and particularly showed women how they could be assertive and not aggressive (Ellis, 1954a, 1954c, 1955b, 1960a, 1963a, 1963b).
5. I was one of the few psychologists who strongly advocated gay liberation in the early 1950s and was made an honorary member of the Mattachine Society (Ellis, 1951a, 1951b, 1952c, 1954a).
6. Along with Kinsey and against Freud and his orthodox followers, I disputed the sacredness of the so-called vaginal orgasm in women and showed both sexes how women could achieve satisfactory noncoital orgasm and still be very healthy and "normal" (Ellis, 1951a, 1951b, 1953a, 1953b, 1953c, 1954a, 1954b).
7. I originated the idea of establishing the Society for the Scientific Study of Sex in 1950, but at first failed to enlist enough support for it. Persisting, and with the aid of Hans Lehfeldt, Robert Sherwin, Harry Benjamin, and Henry Guze, I actually got it going a few years later. It now flourishes as the Society for the Scientific Study of Sexuality.

I am proud of these sexological advocacies and accomplishments. But I was hardly alone in fighting for them, as I had other solid sexologists with me—such as Harry Benjamin (1966; Benjamin and Ellis, 1954), Kelly (1953), and Alfred Kinsey (Kinsey, Pomeroy, and Martin, 1948; Kinsey, Pomeroy, Martin, and Gebhard, 1953). All of us, and many other researchers and clinicians, are steadily making the field of sexology a highly vital and respectable area of science. Let us continue our healthy efforts!

References

Benjamin, H. 1966. *The Transsexual Phenomenon*. New York: Julian.
Benjamin, H., and A. Ellis. 1954. "An objective examination of prostitution." *International Journal of Sexology* 8: 100–105.
Ellis, A. 1945. "The sexual psychology of human hermaphrodites." *Psychosomatic Medicine* 7: 108–25.
———. 1947a. "A comparison of the use of direct and indirect phrasing in personality questionnaires." *Psychological Monographs* 61: 1–41.
———. 1947b. "Questionnaire versus interview methods in the study of human love relationships." *American Sociological Review* 12: 541–43.
———. 1948. "Questionnaire versus interview methods in the study of human love relationships. II. Uncategorized responses." *American Sociological Review* 13: 62–65.
———. 1949a. "Some significant correlations of love and family behavior." *Journal of Social Psychology* 15: 61–76.
———. 1949b. "A study of human love relationships." *Journal of Genetic Psychology* 15: 61–76.
———. 1949c. "A study of the love emotions of American college girls." *International Journal of Sexology* 3: 15–21.
———. 1950. "Love and family relationships of American college girls." *American Journal of Sociology* 55: 550–58.
———. 1951a. *The folklore of sex*. New York: Boni/Doubleday.
———. 1951b. Introduction to D. W. Cory, *The homosexual in America*, pp. ix–xi. New York: Greenberg.
Ellis, A. 1952a. "Applications of clinical psychology to sexual disorders." In *Progress in clinical psychology*, ed. D. Brower and L. A. Abt, vol. I, pp. 467–80. New York: Gruwe & Stratton.
———. 1952b. "Perversions and neurosis." *International Journal of Sexology* 6: 232–33.
———. 1952c. "What is normal sex behavior?" *Complex* 8: 41–51.
———. 1953a. Discussion of W. Stokes and D. Mace, "Premarital sexual behavior." *Marriage and Family Living* 15: 248–49.
———. 1953b. "Is the vaginal orgasm a myth?" In *Sex, society and the individual*, ed. A. P. Pillay and A. Ellis, pp. 155–62. Bombay: International Journal of Sexology Press.
———. 1953c. "Marriage counseling with couples indicating sexual incompatibility." *Marriage and Family Living* 13: 53–59.
———. 1953d. "Recent studies on the sex and love relations of young girls." *International Journal of Sexology* 6: 161–63.

———. 1954a. *The American sexual tragedy.* New York: Twayne. Rev. ed. New York: Lyle Stuart and Grove Press, 1962.

———. 1954b. "Psychosexual and marital problems." In *An introduction to clinical psychology,* ed. L. A. Pennington and I. A. Berg, pp. 264–83. New York: Ronald.

———. ed. 1954c. *Sex life of the American and the Kinsey report.* New York: Greenberg.

———. 1955a. "Masturbation." *Journal of Social Therapy* 1, no. 3: 141–43.

———. 1955b. "Woman as sex aggressor." *Best Years* 1, no. 3: 25–29.

———. 1956a. "New light on masturbation." *The Independent,* Issue 51, 4.

———. 1956b. "On premarital sex relations." *The Independent,* Issue 58, 6.

———. 1957. *How to live with a neurotic: At home and at work.* New York: Crown. Rev. ed. Hollywood, Calif.: Wilshire, 1975.

———. 1958. *Sex without guilt.* New York: Lyle Stuart and Grove Press. Rev. ed. 1966.

———. 1960a. *The art and science of love.* New York: Lyle Stuart and Dell.

———. 1960b. "A case for polygamy." *Nugget* 5, no. 1: 19, 24, 26.

———. 1962. *Reason and emotion in psychotherapy.* Secaucus, NJ.: Citadel.

———. 1963a. *The intelligent woman's guide to manhunting.* New York: Lyle Stuart and Dell Publishing. Rev. ed. *The intelligent woman's guide to dating and mating.* NJ.: Lyle Stuart, 1979.

———. 1963b. *Sex and the single man.* New York: Lyle Stuart and Dell.

———. 1965. *The case for sexual liberty.* Tucson, Ariz.: Seymour Press.

———. 1976. *Sex and the liberated man.* Secaucus, NJ.: Lyle Stuart.

———. 1988. *How to stubbornly refuse to make yourself miserable about anything—yes, anything!* Secaucus, NJ.: Lyle Stuart.

———. 1991. "Achieving self-actualization." *Journal of Social Behavior and Personality* 6, no. 5: 1–18. Reprint New York: Institute for Rational-Emotive Therapy.

———. 1994. *Reason and emotion in psychotherapy.* Revised and updated. New York: Birch Lane Press.

———. 1996. *Better, deeper and more enduring brief therapy.* New York: Brunner/Mazel.

Ellis, A., and R. A. Harper. 1961. *A new guide to rational living.* North Hollywood, Calif.: Wilshire.

———. 1975. *A new guide to rational living.* North Hollywood, Calif.: Wilshire.

Ellis, A., R. A. Harper, S. Dyer, R. Timmons, R. Hill, and N. Kavinoky. 1952. "Premarital sex relations." *Marriage and Family Living* 14: 229–36.

Ellis, A., and E. Sagarin. 1964. *Nymphomania: A study of the oversexed woman.* New York: Gilbert Press and MacFadden-Bartell.

Ellis, H. 1936. *Studies in the psychology of sex.* 2 vols. New York: Random House.

Freud, S. 1965. *Standard edition of the complete psychological works of Sigmund Freud.* New York: Basic Books.

Kelly, G. L. 1953. *Sex manual for those married or about to be.* Augusta, Ga.: Southern Medical Supply Company.

Kinsey, A. C., W. B. Pomeroy, and C. E. Martin. 1948. *Sexual behavior in the human male.* Philadelphia: Saunders.

Kinsey, A. C., W. B. Pomeroy, C. E. Martin, and P. H. Gebhard. 1953. *Sexual behavior in the human female.* Philadelphia Saunders.

LoPiccolo, J., R. Stewart, and B. Watkins. 1972. "Treatment of erectile failure and ejaculatory incompetence with homosexual etiology." *Behavior Therapy* 3: 1–4.
LoPiccolo, J., and L. LoPiccolo, eds. 1978. *Handbook of sex therapy*. New York: Plenum.
Masters, W. H., and V. E. Johnson. 1966. *Human sexual response*. Boston: Little, Brown.
———. 1970. *Human sexual inadequacy*. Boston: Little, Brown.
Pillay, A. R., and A. Ellis. 1953. *Sex, society and the individual*. Bombay: International Journal of Sexology.

Note

This appendix is from Bonnie Bullough, Vern L. Bullough, Marilyn A. Fithian, William E. Hartman, and Randy Sue Klein, eds., *How I Got into Sex* (Amherst, N.Y.: Prometheus, 1997), 131–140. Copyright © 1997. Reprinted by permission of the publisher.

My Path to Sexual Science

IRA L. REISS

Learning to Tell a House from a Home

ONE OF THE MOST SIGNIFICANT gifts given to me by my parents was their acceptance of my personal choices. My mother, in particular, supported everyone's right to challenge the choices society imposed. She was an iconoclast who felt betrayed by a world that permitted so much injustice and unhappiness in our lives. My father was more of a pragmatist, and less confrontational, but he had high regard for intellectual questioning and the search for new knowledge. He was more willing than my mother to adapt to the world as it was, and then quietly and privately make what adjustments were needed. My mother was more like Don Quixote, compelled to battle, even though deep down she knew she could never really win. But, her efforts were not without value, for as George Bernard Shaw has noted, reasonable people conform, and so all progress depends on the unreasonable person.

The city I grew up in afforded me an ideal laboratory to develop my own ideas about sexuality. Scranton was an anthracite coal mining town in northeastern Pennsylvania. My father moved there from New York City in the early 1930s, shortly after my sister and I had started school. He was trying to rebuild the clothing factory he lost in the early years of the depression. Scranton, like most cities at that time, was in deep economic trouble and we lived close to poverty for a few years until finally my father's persistent struggles paid off and he was able to establish a new clothing factory.

At the same time, other entrepreneurial souls in Scranton were founding a different sort of business as their way of surviving the depression years. Houses of prostitution opened—scores of them—in the alleys behind twelve of the downtown streets. The impact on Scranton economically and culturally was dramatic. Prostitution brought in a large amount of wealth to many parts of the city and as a result the sex industry gained considerable political support and it thrived throughout the depression years. On a Saturday night, the population of Scranton increased tremendously. Scores of cars came in from New York City and Philadelphia and many other smaller cities, loaded with men eager to "check out the Scranton girls."

Sexual services from black and mulatto prostitutes were less expensive than the same services from a white prostitute. The cultural script on race, class, and gender for the 1930s and early 1940s was indelibly written in the social and economic structure of Scranton's houses of prostitution.

The houses brought money into the eager cash registers of the hotels, restaurants, gas stations, and bars. As a young boy, even I benefited economically from their existence. No, I did not become a child-pimp or a child-performer. My part-time role was more indirect than that. It was impossible to grow up in Scranton and not know where these brightly painted "pleasure houses" were located. I recall that when I was about ten years old, cars would stop by me when I was walking with friends near the downtown area. The male passengers would lean out the car window and say: "Hey kid, where are the cat houses?" When I gave them directions, the men in the car would usually toss me a nickel or a dime. I was fascinated by the power of sex to so easily bring so many men, so far from their homes, into an economically depressed coal mining town. It reinforced the idea in me that there must be something very special about sexuality—something mysterious, powerful, and extraordinarily rewarding.

When I was in high school, I became a customer and at times I would stay in the parlor after having sex to talk and get to know the "working girls" better. Many of them were high-school aged, about sixteen to eighteen. They would make it clear in their conversations with each other and with me that they were not going to be like the friends or sisters they knew who worked for years in some factory sweat shop for low pay, with no future. They also did some things to protect themselves and their customers. They offered condoms to all who would use them and that started me on the road to safe sex early on. Also, they were regularly inspected for disease by medical people. Many hoped to earn good money, save it, open up a beauty shop or dress shop, and eventually get married. I saw some of these same women years later, and a few had achieved that goal. But many of them found it wasn't such an easy path and left or headed downhill.

Bear in mind that a good deal of the condemnation of prostitution was not from people who supported equality in society for women. The rejection of sex without affection and particularly a negative view of female casual sexuality underlay much of the conservative criticism of prostitution. Some of today's feminist opponents of prostitution have similar views and would not accept prostitution even in a fully gender equal society. Much of this antiprostitution view is based more on a narrow view of what is acceptable sexuality than on a striving for gender equality. In my mind, the harm in prostitution is not in the recreational approach to sexuality, but in the gender inequality and the illegality that still permeates that profession.

Mixing in the Religious Script

Although my parents did not live strictly according to Orthodox Jewish rules, they wanted me to be exposed to that way of thinking. Both my mother and father had parents who were very strict in their religious observances. In the early 1900s, my father's mother, Rachel Goldworm Reiss, helped found a female rights movement within orthodoxy. She was an educated woman who in addition to speaking English and Yiddish could speak Hebrew, the language of the Torah. This ability was a rarity even among religiously educated Jewish men and it gave my grandmother more influence in her efforts to equalize women's rights in Judaism. She was a very determined woman, just like my mother was, but unlike my mother, she had the good fortune of having found a cause to which she could commit her energies.

Every day after public school, until I was Bar Mitzvah-ed, I was sent for two hours to attend an Orthodox Jewish religious school. Every Saturday morning I would attend a four-hour religious service. I wasn't too enthusiastic about this extra schooling but, like it or not, that religious training did have a profound impact on my views concerning sexuality. Over time, the religious school education focused more and more on interpreting the meaning of the Bible, which in Judaism is the Old Testament. We studied the Torah, the five books of Moses, line by line, word by word. We started with the assumption that the Torah was the word of God, but we were all obligated to examine it very carefully and argue the case for different interpretations of that holy script. I did not know it then, but I was learning a respect for logic and reasoning and careful examination of the deeper significance of human belief and behavior. That training was to serve me well in my professional work long after I gave up on Orthodox Judaism.

But along with the respect for logic and reasoning of my religious education came the value assumption that we ought to avoid sexuality outside of marriage. This was at the same time that I was giving directions to out-of-towners as to where the cat houses were located! The religious restrictions included masturbation as well as petting and intercourse and so there was not much sexuality that was acceptable. This religious perspective inserted a third element in my thinking: I had a pleasure emphasis from my mother; a pragmatic emphasis from my father; and now, from my religion, an ethical emphasis that promoted guilt and shame. I was receiving, in spades, a typical conflicted American sexual socialization.

The freedom my parents gave me to work things out coupled with the restraints of my religious upbringing, encouraged me to search for a way to put these opposing pieces of my sexual puzzle into a meaningful guide for living. One of the first areas that I questioned was the double standard that afforded men more sexual rights than women. I reasoned that the double standard restraints on women's sexuality inevitably led to restraints on the range of partner choices that men had, and so self-interest alone spoke against the double standard. My gender equality attitudes were supported by the fact that my mother was a very assertive woman who did not take a back seat to anyone. She also was a woman who let me know that she felt that part of the reason the world had shortchanged her was simply because she was a woman. Furthermore, my sister, Carol, who was just sixteen months older than I, was honored as one of the two best students in her high school class. As I've mentioned, my paternal grandmother had been an active Jewish feminist and had fought in the early 1900s against special male privileges in orthodox religion. With all that in my background, how could I accept any theory of inherent female inferiority and male dominance? I felt lucky to be able to endorse male equality!

It interested me that many of the men I knew were unaware of any inequity or hardship imposed by the double standard on women. The argument they often used was: "Women like it that way" or as Marilyn Quayle put it at the 1992 Republican convention: "Most women do not want to be liberated from their essential natures as women." That kind of explanation did not cut much ice with me. The fact that some women accepted male dominance didn't make it right anymore than some slaves accepting slavery made that right. Furthermore, as I've said, I knew many women who, even in the 1940s, outright rejected such inequality and the reasoning underlying it.

A Sex Educator in the Army

Another important influence on my thoughts and feelings regarding sexuality came from my hitch in the U.S. Army (1944–1946). I turned eighteen in December 1943 and was drafted just weeks later. After army testing, I was placed into a unit that was supposed to be sent to college for special training. Many of the men in this outfit already had started college when they were drafted and many came from upper-middle-class homes. Very few came from coal mining, open-prostitution towns like Scranton. What most surprised me about these men was that, although they were about eighteen to twenty-two years old, many of them were still virginal or had very limited sexual experiences. I had thought the world was like Scranton where very few boys passed the age of eighteen without sexual experience—but I soon learned otherwise. The army changed its plans from sending us to college for special training and instead sent most of us for basic training as part of a combat engineer battalion at Camp Chafee, Arkansas. The army taught us how to build roads and bridges and then how to blow them up when we were done. Then they sent us overseas to try out our skills against the Nazis. We went to England for more training and then landed in Normandy.

Many of my army buddies saw that I was more comfortable with women than they were. I saw in these better-educated men a chance to learn about things they knew more about, like good books, a wider range of music, how to play chess, and much more. As things worked out, we made a fair exchange; they picked up from me ways of meeting and getting involved with women and I learned from them some of what they knew about literature, music, politics, and more. I think we all benefited from the exchange. My army experience convinced me more than anything that biology is flexible and is not the major determinant of one's sexual attitudes and behaviors. These army friends were no different than my Scranton friends in their biological inheritance, but they were a world apart in sexual attitudes and behavior.

Climbing up the Academic Tree

After the army I went to Syracuse University and in 1948 took a philosophy course that was particularly important in moving me in my career direction. My professor was critical of all premarital sex because he had witnessed how it could lead to pregnancy and disease. I rejected that global negative view because safe sex had worked very well for me in avoiding disease and pregnancy outcomes. I argued my ideas with him in class and in the written work I did for that course. I continued the discussion of this issue in depth with the six college roommates with whom I shared the second floor of a home near campus. In order to try to convince my roommates about the worth of my ideas, I wrote a paper on choices among premarital sexual standards. This was my first written attempt at integrating the conflicting elements that existed in my own sexual socialization. In this paper, I presented the conventional view of marriage as the ideal place for a sexual relationship but I suggested that it was also acceptable to have sex under other conditions. I stressed the value of affectionate stable sexual relationships but I accepted recreational "body-centered" sex as well. I felt that one could enjoy recreational sex, but it was best to make it a secondary and not a primary activity lest you miss out on the greater rewards of affectionate sexuality.

I rejected abstinence as *the* answer. I stressed the importance of the spirit and the intention in a relationship rather than following the strict letter of some demanding orthodoxy. I saw virginity as a mental state and not a physical state. Each sexual relationship actualized something very new in the world and so the fact of having had prior sexual relationships was not very important. In every new relationship, both partners were in a psychic sense virginal. I put forth these personal views equally for both men and women and rejected the double standard.

I was taking a personal position in accord with the standard I would later call "permissiveness with affection" and that I would herald as the wave of the future. The satisfaction I gained in writing this 1948 paper and arguing about it with my roommates encouraged me to plan to someday write a book on premarital sexual standards. Clearly, I was working out my own personal problems, but equally so, I saw that my problems reflected the cultural and social confusion regarding sexuality that has disturbed so many of our people for so long a time.

The Choice of a Career

After college, I went to work for my father in his factory. But I was not happy doing that and after a year I decided to find a profession that would better fit my intellectual interests. But I wasn't sure what that might be. As my unhappiness at work increased, I discussed my feelings with Mort Friedman, my old high school friend who was also one of my Syracuse roommates. I always liked to argue and debate controversial issues and one night after work Mort and I were having our usual debate about some social issue and Mort sighed, paused, and turned to me and said: "You know Ira, you really enjoy battling over these controversial issues. Maybe you ought to be a college professor and get paid for doing that." This was a landmark experience. Mort's suggestion was so much in harmony with my gut feelings that I knew my life was about to take a dramatic turn.

I began to think intensely about going back to graduate school and getting a doctorate and becoming a college professor, but I wasn't sure what field to choose. My first thought was philosophy for I loved the open questioning of that field, but I still had my old feeling that philosophy left too much up in the air because it lacked an established tradition of empirical verification. I wanted a field that dealt with philosophical issues but that used scientific methods to help resolve differences. I also liked psychology, but I felt it was too morbid. Another college friend of mine helped clarify things for me. Ralph Forest had taken sociology and he told me that he believed sociology was just right for my type of interests and advised me to get my doctorate in that field. I still wasn't fully certain, but I made up my mind that I would start my graduate work in sociology but take courses in philosophy too and decide which one to major in during my first year in graduate school. My parents backed up my decision and I entered Pennsylvania State University in the fall of 1950 to work toward my doctorate.

Academic Hurdles and the Race to the Finish Line

The sociology department at Penn State was relatively new and had given its first doctorate just four years earlier, in 1946. The recipient was William J. Goode who years later was to make his mark in sociology and eventually become president of the American Sociological Association. My first sociology course decided my career choice for me. Luther Lee Bernard, a former president of the American Sociological Society, was my professor. He was sixty-nine years old and this was to be the last course of his life, but he exuded excitement about sociology and it was contagious. A few weeks into the semester I gave my first graduate report (on Egyptian religion) to Bernard's class. I can still see him staring at me with his blue eyes gleaming with approval during my talk. My fellow grad students also responded with enthusiasm. Afterwards, they followed me out of the class, asking me how I knew so much about Egyptian religion and how was I able to conceptualize it so clearly. I was thrilled. I had only read one book. I thought to myself: "Imagine what I could do if I really knew the area I was reporting on!"

This very positive reaction gave me the confidence that I could organize ideas into coherent patterns and make them interesting and understandable. I knew then that I could be a teacher and perhaps a writer and I could contribute something to sociology. Sociology was now my chosen major area. Philosophy would become my minor area for my doctorate. The fire to learn was fully lit and I was eager to obtain my degree and start my career. I had found a pathway into the academy and I was anxious to fully explore what lay ahead.

By far the most influential professor for me at Penn State was a young, imaginative, free-thinking sociologist named Edward J. Abramson. I took his course in social change and I liked the breadth of his qualitative view of sociology, the paradoxes he saw in society, and the personal respect he gave to even my sometimes far out ideas about society. As a newcomer to the field, I needed that sort of support, even though I covered that need up with my brash presentation of self.

Another very influential person in my first year was William Bensch, a fellow graduate student. He had just returned from a year's trip around the world and he was always willing to argue and discuss new ideas. We often disagreed but we also clearly showed the mutual respect that gave the lie to the ridicule we hurled at each others' views. Bensch, like Abramson, had the depth of cultural background and tolerance of difference to which I was attracted. I knew I had a lot to learn, but I also was not about to give up my right to decide what I accepted. I always questioned things before I would accept them, even such widely accepted things as the poetic value of William Shakespeare or the power of Wagnerian music. Finding acceptance for my feisty approach to learning enabled me to continue my cultural education.

My first and most lasting love in sociology was the development of theories that make sense out of some part of our social life. And so theory construction became my major area for my doctorate. I was fascinated to find out how things worked in society and why things worked the way they did, especially in areas of controversy, like sexuality. I minored in cultural anthropology for my master's and minored in philosophy of ethics for my doctorate. Both of these minors opened up windows into the universe of ideas that I eagerly wanted to explore. The years in graduate school were an intellectual feast for me and I gorged myself as much as I could. But I wanted to be a professor and be free of other people's rules and regulations. I rushed through, earning my master's and doctorate degrees and making up the undergraduate sociology courses in just three years.

In the second of those three years, I went to Columbia University to take my doctorate course work and gain a different perspective. Once there I was very impressed by the theoretical abilities and interests of Robert K. Merton. He was only forty-one then and four years later he was to become president of the American Sociological Association. Most importantly, he was a very stimulating and exciting scholar. Unlike Abramson and Bensch, Merton was not that approachable and was far more systematic and scientific in his thinking. Nevertheless, his vast knowledge and ability to explain social events was exceptional. He influenced me to accept the worth of many of the ideas of the structural/functional approach and to put more emphasis on the scientific aspects of sociology.

In my third and final year, I went back to Penn State to take my language and preliminary exams and write my dissertation. What happened with my dissertation project was quite anxiety provoking and revealed a great deal to me about academic life. My dissertation was an ambitious project aimed at comparing sociological viewpoints concerning how to obtain the subjective thoughts and feelings of people in society. I compared the qualitative versus quantitative methodological positions on this issue and developed my own perspective. This issue held great interest for me and was part of what was then a widespread debate about just how scientific sociology could be in studying people's subjective beliefs. This dissertation combined my interest in theoretical explanation with my interest in controversy.

A heavy dose of departmental politics interjected itself into my dissertation project. Walter Coutu, one of the faculty members on my committee, thought he had written the answer to my dissertation inquiry in his book (Coutu 1949). I did mention his ideas in my first dissertation draft but I had not given them prominence, nor had I endorsed them. I strongly believed in my right to academic freedom, but I was soon to find out how academic politics can compromise academic freedom, particularly if you are a graduate student.

Arnold Green, my Ph.D. advisor, was an associate professor. Coutu was a full professor who would vote on Green when he came up for promotion. This worried Green and he told me that he hesitated to set up my final oral while my conflict with Coutu was unresolved. The heart of my dissertation dealt with whether in the study of people's viewpoints you gave priority to the meaning and significance of people's viewpoints or whether you gave priority to the reliability of evidence concerning those viewpoints. I felt that the significance of what was being studied about people's subjective viewpoints was of prime importance even if it could not easily be replicated. I did not disregard the value of reliability but I thought of it as secondary in importance. My position was in conflict with Coutu's views. I consulted with Seth Russell, another committee member, and he advised me to take a more middle-of-the-road position in my final dissertation draft and afford significance and reliability a more equal share of importance. In the interests of finishing up in time to take my first job, I modified the final draft of the dissertation.

But that was not enough to end the conflict. Coutu came down more on the side of reliability than I did and he thought his perspective resolved the controversy. Of course, I discussed his view in my final draft but did not describe it as having resolved the controversy. I had compromised as far as I was willing to go. But I had only a few weeks left in which to schedule my final oral before leaving for my first job, and Green was still unwilling to schedule it. I was getting desperate but then I thought of a way out of this dilemma.

Since Green respected rank, I decided I needed to somehow position rank on my side. Russell was the chair of the department and was therefore the highest ranking person on my committee. I could not talk directly to Russell about Green's hesitancy to set up my final oral without aggravating the interdepartmental conflict further and so I took another path. I called Russell and told him that Green and I had talked about having my oral on August 21. I asked him, as chair of the department, for his approval to finalize this date. Russell thought my statement about talking with Green meant that Green had agreed to set up my oral. That was what I hoped my vagueness would accomplish. Russell endorsed the date for my final oral. It was true that I had talked about that date with Green. I simply neglected to add that Green had not agreed to schedule it then. Desperate times call for desperate measures.

After I spoke to Russell, I proceeded to call Green and told him that Russell wanted my final oral to be scheduled on the twenty-first. I knew that Green respected rank and therefore he would not deny what he thought were the chair's wishes. Also, by Russell making the decision, Green knew he could not be criticized by Coutu. Green agreed with setting up the oral. I then called Coutu and told him that Russell and Green had agreed to set up the oral on the twenty-first.

He did not object. The rest of my committee all accepted the date. I had played the cards in the power game and won that opening round. But what would happen at the oral was still up for grabs.

My oral was scheduled at 3:30 on Friday, August 21. It was the last day possible for holding an oral before the fall semester. I had to be finished this day if I was to get my degree before leaving for my first job. The oral began with the usual formalities. The difference of opinion with Coutu came out during the oral but in muted form. Coutu was not as critical as I had feared. Also, Green sought to shift the discussion to areas that would not arouse Coutu's disagreement. Russell tried to reduce the tension by suggesting minor changes that should be made in the dissertation. Russell had shown support for me throughout my stay at Penn State and he provided the needed leadership that afternoon to keep the oral on track toward a successful completion. During the oral, the other four members of the committee expressed support for my dissertation. Then the discussion ran its course and Green asked me to leave the room so they could decide if I passed.

When I went out in the hall to wait the final decision of the committee, I knew the start of my career was in jeopardy and I was very unsure of the outcome. The committee could have decided to ask me to radically rewrite and thereby delay my degree a full year. I very much wanted to start my first job without that burden. Finally, Green called me back into the room and informed me that although I had to make a number of minor changes in the dissertation, I had passed. He did not express much in the way of congratulations but informed me in a matter-of-fact fashion. But being passed was the only outcome that mattered to me. I don't know how Coutu voted but I do believe he voted to pass me because he came to the graduate student party that my friends had that night and personally congratulated me. So perhaps the problem on my dissertation was present more in Green's anxiety about promotion than with Coutu's disagreement with my dissertation. This experience was an early lesson in the place of power in the realm of ideas. There surely was and still is an adversarial dimension to academic freedom.

My Professional World: The Early Years

A few weeks later, in September 1953, I started in my first job at Bowdoin College in Maine. Alfred Kinsey had received his bachelor's degree there thirty some years earlier. I learned that the college was thinking of giving Kinsey an honorary degree but they changed their mind for fear of criticism from Senator Joseph McCarthy who was then spewing his bigotry even into the colleges and universities of our country.

Burton Taylor, the chair of my department at Bowdoin, asked me to teach a family course. I did not know it then, but that course was to afford me my first legitimate platform for expressing my ideas about gender and sexuality. Having never taken a family course, I wrote away for all the texts in that area and proceeded to try to select one. I felt I knew something about sexuality—after all, I was the unofficial sex educator of the 1273rd Combat Engineer Battalion and the author of a 1948 college position paper. I decided I would read the chapter on sexuality in each textbook and choose on that basis. What I found was shocking.

Almost without exception every textbook in the family area condemned premarital sexual intercourse. Sex before marriage was painted as inevitably involving negative outcomes like pregnancy, disease, guilt, social condemnation, and the weakening of your future marriage. Your relationship might escape one or two of these, however, you could never escape all of them. The positive consequences of premarital sex like physical pleasure, psychological satisfaction, or preparation for marital sexuality were ignored or distorted. In general, pleasure outside of marital sex was conceptualized as evil and lustful. In the perspective of these textbooks, there was no affectionate basis for a premarital sexual relationship and selfishness and irresponsibility were the key motives.

Of course, some premarital sex fit the portrait being sketched in these textbooks, but there was so much more that was left out or misrepresented. To accept what was said in these texts was equivalent to accepting as accurate a politician's description of his opponent. These textbooks' treatment of sexuality were not scientific, nor were they even carefully reasoned. Data were either ignored or distorted. The textbooks were simply boldface traditional moral propaganda.

I was taken aback by the presentation of such dogma as social science. One of my major goals in my career was set by this experience. I vowed that I would present a more empirically based and logically reasoned view of sexuality in my family class and in my writings. Also, I would work hard to not allow my values to bias my presentation in my publications. I could not eliminate my values, but I could work to keep them from blinding me as to what the world was like. Some of my early hesitancy to take a strong policy stance as part of my sociological analysis was tied to the destructive mix of value positions and sociology that I had found in those family textbooks. It took time for me to learn how to integrate my values and my policy suggestions into my work and still maintain the fairness and balance of the empirical work.

Just as I was writing my first two articles (Reiss 1956, 1957) criticizing the family textbooks I had examined, I met the woman who has been my life companion every since: Harriet Marilyn Eisman. She has been my intellectual sparing partner as well as my editor and emotional support since I met her in late January 1955, fell in love with her, and married her in early September 1955. She discussed all my writings with me and gave me much to think about. But it was only in my 1990 book that she allowed me to even list her name on the cover of the book. So as I explain my academic career, I want to here give major credit to the wonderful collaboration that she and I have had on all my work since our marriage in 1955. She combined the finest traits of both my parents: the complete devotion of my mother and the pragmatism of my father. Having Harriet in my life has made everything else in my life much more enjoyable.

Just weeks after we married, I began a position at the College of William and Mary in Williamsburg, Virginia. I wrote the early drafts of my 1960 book on premarital sexual standards in the summers of 1956, 1957, and 1958. It was in that book, in the last chapter, that I predicted the sexual revolution that I believed was to fully appear in the late 1960s (1960a, 239–241). William Kephart at the University of Pennsylvania reviewed that book in sociology's top journal *The American Sociological Review* in April 1961. He commented on my prediction of college students leading a new sexual revolution by saying that in "the last chapter . . . Reiss apparently lost control of the typewriter" (1961, 294). But my prediction was exactly the way our country went—so we best watch out for book reviewers who lose control of their typewriter.

I saw the major cultural trend among young people in our country to be a rejection of abstinence as "the one and only" sexual standard and as a questioning of the fairness of the double standard. The trend, as I saw it, was an endorsement of permissiveness with affection as an acceptable choice in place of these two ancient standards. In connection with writing about this new affection-based standard, I developed my ideas about how love develops in dyadic relationships and presented them in the form of my "wheel theory of love," which has over the years been one of my most popular articles (Reiss 1960a, 136–144; 1960b). My book also spoke about permissiveness without affection and of some increases in that standard. But the dominant values in American society did not support sex without affection anywhere near as much as sex with affection. While it was true that I preferred the trends that I was reporting, I did analyze my prediction using the best data we had on cultural trends and worked hard to be as fair and unbiased as I could. The fact that permissiveness with affection did indeed become the most popular new sexual standard supported the fairness and accuracy of my empirical analysis and my theoretical interpretation of the changes in our society.

It was my questioning of the relative value of sex without affection that led me to correspond with Albert Ellis who was one of the few writers who supported a broader range of sexuality than I did. I wrote to Ellis about my reservations concerning the strong support for casual sex that he expressed in some of his writings. We corresponded and discovered that we were not quite as far apart as we thought and our correspondence and friendship built from that point forward. He is the person who first moved to found the Society for the Scientific Study of Sexuality. He convinced me to become a charter member of that organization when it began in 1957. Ellis is still a leader in confronting the dogmatism of the radical right. All of us who value choice in sexuality owe him a debt of gratitude for the many battles he has fought—and won.

The years at William and Mary rounded out my basic background in sociology and also in anthropology. There were only four of us in the department and so we each had to teach many different courses. We required a senior research paper for sociology majors. In 1958, with the encouragement of my chair Wayne Kernodle I started a major study of premarital sexual standards. It was my way of testing the idea that many young people do indeed believe in newer sexual standards such as the one I called "permissiveness with affection." It was the increasing popularity of this new standard that I felt heralded a new sexual revolution in our country. Four of our senior majors chose to work with me on my project: Ron Dusek, Martha Fisher, Richard Shirey, and John Stephenson. Stephenson was the leader of the group and the one who would go on to get a Ph.D. in sociology and become president of Berea College in Kentucky. All four students worked extremely long and hard during their entire senior year on this project. They had been caught up in the passion of this project and they were excited about the possible findings.

I discussed the development of my Guttman scales to measure premarital sexual permissiveness with this group of four students and informed them that their research would be its first testing ground. The twelve-item scales measured premarital sexual permissiveness attitudes by asking about the acceptance of kissing, petting, and coital behavior under conditions of various degrees of affection from "no affection" to "in love and engaged." This scale in both its original and in its shortened form has been used in scores of other research projects (Reiss 1964a, 1967, 1998; Reiss and Miller 1979; Schwartz and Reiss 1995). But to test these scales and our hypotheses concerning causes of changes in the permissiveness levels they measure, we needed to get permission from local high schools and colleges to administer our questionnaire to their students. In Virginia, in 1958, this was not an easy thing to do. When my four senior students approached the principals of the white and the black high schools in Williamsburg, they were turned down.

I knew that these principals feared public reaction, but I also hoped that they would see the usefulness of knowing more about their students. I had to figure out how to change the priority of these two outcomes in their thinking. I went to see the white principal first and when he repeated his opposition to administering my questionnaire, I asked him how many of his high school girls had become pregnant the year before. I then asked him whether he felt any responsibility to lessen the risk of pregnancy for these and for future girls. I also said to him: "Do you want to be seen as someone who chose to pass up the opportunity for increasing awareness of sexual standards that might help reduce unwanted pregnancy?" I stressed that this research was important to gaining knowledge concerning how to avoid outcomes like pregnancy. I also made it clear that I might well make a public issue of it if he continued to deny us permission. He finally came around and said the tenth, eleventh, and twelfth graders could participate but he would not let us give our questionnaire to the ninth graders.

The response from the black high school principal was more of an enigma. Kernodle was friends with this man and he took me to his office and introduced me. The three of us sat down to talk and I explained the importance of this research for understanding our students better and gaining control over unwanted sexual outcomes like pregnancy. But instead of responding, he kept changing the topic to the anti-school-integration politics in Virginia. I kept trying to turn the conversation back to my study and he ignored my efforts. Finally, I realized that he wanted to see where I stood on the still hot issues of racial integration that the powerful Byrd political machine, represented by Governor Almond of Virginia, was resisting. I turned to him and shared a political joke I had heard. I asked him if he heard that the Byrd machine had inbred so much that it finally had produced an idiot—the governor. He gave a loud laugh and said to me: "When do you want to give out the questionnaires?" The portrait of the politics of doing sex research in the 1950s was becoming clearer and clearer.

We gave out our questionnaires to the two Williamsburg high schools and also to two colleges (William and Mary and Hampton Institute). The four students on the project worked with our primitive McBee card system. It was a precomputer method of analysis used to test whether our premarital questions produced a usable Guttman-type scale. The answer came one night in the spring of 1959. I remember that night very well because we lived right across the street from the campus and there was a "panty raid" that night. Harriet and I were observing the event with interest from our window. The phone rang and I thought it must be about the panty raid on campus. But it was John Stephenson and with great emotion he said to me: "They work! They work! The scales work! They meet all Guttman-scale criteria, they are valid!"

The students and I celebrated that event with much gusto. It was like the discovery of some new chemical element, for there was no existing scientific scale to measure premarital sexual permissiveness. This was an important step toward systematizing the analysis of premarital sexuality and getting away from the very unscientific presentations that abounded in the textbooks and in the media. These scales have since come to be very widely used. I have since published a four-item version that can be used instead of the two twelve-item male and female scales (Reiss 1998). This new and very short version has been tested in the United States and Sweden and found to meet Guttman scale requirements (Schwartz and Reiss

1995). The four-item scale is more convenient and one could even use just a single item from that scale if questionnaire space was at a premium (question 3).

In 1959, I left William and Mary and went to Bard College. I applied for a National Institutes of Mental Health (NIMH) government grant to support analysis of my research data. In 1960, I received approval of the first of three NIMH grants that for four years were to support my analyses of these data and my gathering of a new nationally representative sample of adults to further test the scales and the sociological predictors of scale values. When I was awarded the grant, an official from the NIMH told me to change the title of my grant so that it did not contain the word "sexual," or else the grant might not get through all the official government checks. Even today, grants researching sexuality run high risks. It is sad to see how powerful the uninformed and narrow minded still are in our political system.

My new research grants tested out many of the ideas in my 1960 book. That book sought to explain sexual standards in America and predicted dramatic sexual changes by the end of the 1960s (Reiss 1960a, 239–241). The notion that someone could fairly analyze our society and accurately predict such changes was not widespread at that time. I hoped that my research grants would yield knowledge that would dissuade many of such pessimistic beliefs. A job offer from the University of Iowa followed very soon after the book appeared and I decided that it was time to leave the liberal arts college world and explore the Big Ten universities. I've always valued my three liberal arts college positions—especially William and Mary. But I thought it was time to move on and explore an environment that would provide me with better research support for analyzing the data and other work connected to my NIMH grants.

The University of Iowa afforded me good technical support in the way of staff and equipment. The first day I was on campus I came in contact with Don McTavish, a Ph.D. student in sociology. He was quite knowledgeable about computers and I hired him that same day. He was very helpful in my analysis of data during the next three years that he stayed with me. He put our data on IBM cards that could be used by the mainframes then in existence. In 1962, I applied for and received a supplement to my NIMH grant that allowed me to try out my scales on a nationally representative sample of the country. The National Opinion Research Center (NORC) at the University of Chicago was doing "amalgam" surveys where you could buy a segment of time and have your questions asked of a nationally representative sample. My questions were in the NORC June 1963 national sample. It was the first scale analysis of sexual attitudes done on a representative national sample and today it affords us a benchmark measure of the premarital sexual attitudes in this country just shortly before the sexual revolution exploded forth in the late 1960s. These data are now archived at the Kinsey Institute at Indiana University and are available for use by other researchers. I have recently used these benchmark 1963 data and analyzed trends from other national surveys up to 1998 (Reiss 2001).

In 1967, my book analyzing the national sample and the regional school samples was published. It received a very favorable reception and the "autonomy theory" I put forth to explain change in sexual permissiveness has inspired scores of research in the years since (Reiss 1967; Reiss and Miller 1979; Schwartz and Reiss 1995; Hopkins 2000). As noted, I used a national sample as well as student samples to test my Guttman scales of premarital sexual permissiveness and to analyze a number of explanatory variables derived from my 1960 book and elsewhere. The "autonomy theory" that I developed explained changes in premarital sexual permissiveness as due to increased freedom for youth from adult institutions and increased acceptance of sexuality in the culture at large. This autonomy theory was a broad theoretical integration of the seven theoretical propositions that I developed in this book to explain more specifically the social forces that alter premarital sexual permissiveness levels (Reiss 1967). The events that happened in America in the decade after the publication of this book fit very well with the predictions that the autonomy theory made about causes of changes in sexual customs in our country. For example, the theory predicted that the group whose autonomy (ability to run its own life) increased the most would be the group whose premarital acceptance of sexuality would increase the most. Of the four race/sex groups, white females were showing the greatest increase in autonomy and so I predicted their increase in acceptance of sexuality would be the greatest. That was precisely what happened in the decade from the late 1960s to the late 1970s.

The Move to Minnesota

I was at a crossroads in 1967—what to do next? My two books had attracted a great deal of favorable attention and job offers and other opportunities were coming my way. I chose to write a sociology of the family textbook as my next project. Ever since I saw those family textbooks back in 1953 when I first taught a course on the family, I had wanted to present a sociological analysis of the family in what I believed was a more reasonable and accurate fashion. I had been teaching family courses regularly since I received my degree. In 1965, I published an article arguing that nurturance of the newborn was the only family function that was present in every society (Reiss 1965a). I tied that conception into a deeper understanding of the place of the family in society and I wanted to express my thinking on this in a textbook. I also desired very much to write a text that integrated human sexuality into the understanding of the family more than current family textbooks did.

This family textbook eventually went into four editions (Reiss 1971, 1976, 1980; Reiss and Lee 1988). That textbook writing took up a good deal of my time and energy especially during the 1970s. I sometimes wonder just how different things might have been had I focused more on my work directly in the sexuality field and not ventured so deeply into the sociology of the family area. But I did use my sexuality research and theory perspective in my family textbook. Also during the 1970s, I surely was not spending all my time on the textbook. I consistently moved back to sexuality research and theory projects because that was still clearly my number one substantive interest (Reiss, Banwart, and Foreman 1975; Reiss and Miller 1979; Reiss, Anderson, and Sponaugle 1980).

In January 1969, as I was working on the first draft of my family textbook, the University of Minnesota offered me a position in the sociology department as the director of the Family Study Center. Reuben Hill wanted to step down from that position in the center, which he had helped found. Interestingly, he selected me despite the fact that he knew I had long been critical of some of his views about sexuality (Reiss 1957). I accepted the offer and joined what was probably then the strongest program in family sociology in the country. Oddly enough, Reuben Hill, the conservative Mormon, and Ira Reiss, the liberal Jew, got along very well with each other. We both accepted each other's diverse viewpoints and our strong sense of loyalty to good research and theory helped bind us together during the years we were colleagues at Minnesota. Reuben was the most helpful colleague I had during those years. He and his wife Marion became our friends. Harriet and I were both very grieved by his death in 1985 and by Marion's in 2001.

At the 1971 National Council on Family Relations meeting, Reuben Hill and I together with Ivan Nye and Wesley Burr organized a major project to present all the existing theories in the family field in a formal, scientific fashion in a two-volume work. The book was published in 1979 after eight years of effort to get some of the best people in the family field to write up their formal theories on various substantive areas of the family (Burr et al. 1979). Together with Brent Miller, one of my doctoral students, I wrote a chapter for this book analyzing the recent testing and development of my 1967 autonomy theory (Reiss and Miller 1979). In the 1970s, I was more zeroed into the formalization of theory as the key way to develop our theoretical understanding of the family. Today, I would broaden that approach to include more qualitative data and more policy implications.

The role of director of the Family Study Center at the University of Minnesota stressed administration work and the obtaining of research and fellowship grants for graduate students. I have never liked administering to other people's problems. I stayed as director for five years but halfway through that term I became eager to get out of that role. I wanted to be free to focus more fully on my own research and theory work.

I was once again searching for a new challenge. I went on my first sabbatical in 1975–1976. In order to gain a broader view of how gender and sexuality are tied into different societies, I spent that year teaching and doing research at Uppsala University in Sweden. By being a foreigner for that year, I learned how important it is to understand the cultural and social context of sexuality and gender. It was obvious that the Swedes had a deeper acceptance than we do of premarital intercourse as a legitimate option for young people. The public display of condom advertisements on billboards and their availability in street machines was one indicator of this. The lower disease and pregnancy rates in Sweden made me aware that this greater acceptance and better preparation for sexuality was indeed working very well. I have found my experience in Sweden very helpful in developing my cross-cultural perspective on sexuality, gender, and the family (Reiss 1980).

Part of what I learned about Sweden concerned extramarital sex. Sweden seemed to have high rates of premarital sex but moderate rates of extramarital sex. I decided to examine extramarital sex attitudes in America and search out their best predictors. Together with G. C. Sponaugle, a doctorate student of mine, and my colleague Ron Anderson I undertook to analyze U.S. national data and develop a theory explaining attitudes toward extramarital sexual permissiveness. We used four different years of the General Social Surveys data on extramarital sex. That sample was gathered annually by NORC and was representative of the nation. Although it used only one rather general question to measure extramarital attitudes, it was the best national sample data available. We carefully analyzed some hypotheses about possible predictors of changes in extramarital sexual attitudes. It turned out that there were three major factors influencing extramarital attitudes. I would sum them up as: (1) how happy one's current marriage was; (2) how highly one valued sexuality; and (3) how intellectually flexible one's attitudes were (Reiss, Anderson, and Sponaugle 1980). The findings contradicted the popular, simplistic view that the first factor, unhappiness of marriage, was the key answer to why extramarital sex was accepted. It was not even the strongest predictor. Several replicative studies of our work followed (Glass and Wright 1992; Reiss 1998). I had thought about doing a book on extramarital sexuality, but I felt I had done enough in this major project and I resumed my search for another area to explore.

Coming Full Circle

By the 1980s, trends toward the increased acceptance of sexuality stalled and sex negative resistance increasingly surfaced. I felt the need to speak out more in my publications about the directions that we as a society should be taking and explaining why the current reactionary movement would not resolve our sexual problems. I believed that if I was aware of my own values and made a conscious effort to avoid bias, I could make my value assumptions explicit and still be fair and balanced in analyzing the sexual situation in our country and in proposing solutions. The increased resistance to sexual changes had made me more conscious of the inevitable mix of science and values when we study human behavior of any kind. I also saw more clearly the need for becoming part of the policy scene by going beyond describing and explaining. I increasingly felt that part of my role as a sexual scientist should involve prescribing ways of better containing our many sexual problems.

By 1980, I also decided that I wanted to reach an audience broader than just sociologists. One project that had interested me even back in graduate school was to examine the writings and data on sexuality cross-culturally with the aim of arriving at a sociological explanation that would be useful in understanding sexuality in any society. This sort of cross-cultural explanation I thought would be useful to any social scientist and also to clinicians and social workers who dealt with problem areas of sexuality. I always had a strong cross-cultural interest and had taught anthropology for four years at William and Mary. My goal in this new book project was to find those parts of human societies that are most strongly linked to our sexual customs and to build an explanation of how and why the specifics of those linkages differed in various societies around the world. This cross-cultural project was an immense undertaking. It was to occupy me for almost the next five years.

I read all the anthropological literature I could find on sexuality in various societies. In addition, I used the Standard Cross Cultural Sample (SCCS) of 186 nonindustrial societies to check out a number of my ideas (Murdock and White 1969). I arranged to get permission to add to the SCCS several new sexuality codes that other researchers working with this sample had developed. My project was very broad and there were times I wondered if I would ever be able to finish it. But eventually I developed my PIK theory, which stated that the crucial linkages of sexuality in any society were in the areas of Power, Ideology, and Kinship (PIK). I tried to explain why sexuality was always linked to PIK and how and why the specific nature of this triple linkage differed in various societies (Reiss 1986). I was pleased with the end result because I believe I had presented a broad macrosociological perspective explaining similarities and differences in the sociocultural linkages of sexuality in human societies. Up until that time, sociologists had not put forth such a cross-cultural theoretical explanation of sexuality.

The comparative approach I used was contrary to the fads and fashions, that is, the politically correct view of the day. The popular view then and still now is that each society is unique and thus we cannot really compare different societies in any meaningful way. I accepted the distinctiveness of each society but I rejected the view that comparisons were not possible. We do compare when we examine people. People are certainly different from each other but surely we can compare them and still accept that they differ in some respects. Science is not possible if such comparisons cannot be made. When the book appeared in 1986, it met with surprisingly strong support from some quarters. For example, Ron Moglia at the graduate program in human sexuality in New York University and others wrote that they were using this book in their cross-cultural sexuality courses. But my ideas did not reach as far as I wanted into the broader professional audience. Nevertheless, I have noticed in recent years that this book is being cited by other writers who share my desire to understand the similarities as well as the differences in sexuality around the world (Suggs and Miracle 1993; Deven and Meridith 1997). Perhaps, the politically correct academic stranglehold on cross-cultural comparisons is beginning to weaken. Science thrives best in an open atmosphere rather than in one where certain types of projects are blocked by prejudgments of their worth (Reiss 1999).

I had taken a significant step in this book away from writing only for sociologists. The change was refreshing to me but it increased my appetite for reaching an even broader segment of the professional world and also the educated public. I once more anguished over how to do this and what sexuality topic to explore. As I mentioned earlier, there was a radical right sexual element growing in our society and that more than anything else motivated me to become more prescriptive in my new project. The reactionaries pushed for returning to an "abstinence only" solution for our sexual problems such as HIV/AIDS. I knew that approach was doomed to failure but I also knew that the American public was very poorly informed and still harbored the Victorian virus that could be activated to support a return to older narrow sexual standards. I wanted to do what I could to show the public that the resolution to our sexual crises was not to be found in the past but in a changed future (Reiss and Leik 1989).

Projects that have a broad societal applicability have always been most interesting to me. I see as my strength the ability to interpret and explain very diverse phenomena. But I had not applied that ability to finding solutions for our society's sexual problems quite as directly as I was to do now. I chose the four American sexual problem areas that begged for problem resolutions: HIV/AIDS, rape, teenage pregnancy, and child sexual abuse. I explored what in our society had produced our high rate of these problems and what changes in society would best bring these four unwanted problems under better control.

As my analyses of these four problem areas proceeded and I examined the major studies in these areas, I found myself confirming that the traditional sexual ethic of our society had seriously exacerbated, rather than helped, in all these problems. It was also increasingly clear that a new pluralistic sexual philosophy was evolving that would help us in managing these sexual problems. That new philosophy was what I called HER sexual pluralism. This sexual ethic asserted that if a sexual relationship was Honest, Equal, and Responsible (HER), then it was morally acceptable whether it was between two men, two women, a man and a women, two sixteen year olds, and so on. Also, the amount of affection was not the number-one value. The HER in a relationship were the key moral criteria of sexual choices in that relationship. Love by itself was not the answer to a moral choice. Surely there were many love relationships with severe gaps in HER and nonlove relationships with greater HER. This change toward a HER sexual standard in America in the late 1980s was as significant to me as the change to permissiveness with affection was in the late 1950s. It heralded a new way of conceptualizing sexuality by the American public. This was, I believe, the final phase of the sexual revolution that had begun a generation earlier and it afforded the best guide to containing our major sexual problems.

Writing this more prescriptive book afforded me such an exhilarating feeling. I was finally openly involving much more of myself in my professional work. Now, I was speaking as a whole person and not just from a "pure" scientific position. Now my work was more of a human endeavor and less of an antiseptic science project. Now I could develop a more humane model of science. It was about time!

But I want to be clear about this new sexual standard. HER sexual pluralism does not say anything goes. It demands an ethical approach that shows concern for one's partner. That concern for partners is the minimal price of admission to the arena of ethical sexual choices. What choices one makes within a HER negotiation will vary by where one is in his or her life and what other values one holds. This minimalist HER philosophy would encourage people to take responsibility for working out such ethical choices and to be open to changing their specific preferences when their life situations and values change. But not all choices are acceptable. HER pluralism does not require a commitment to deep affection, but it does require a concern for treating the other person in an honest, equal, and responsible fashion. Such a relationship would have to avoid the use of force and exploitation. HER sexual pluralism surely qualifies as a revolutionary perspective compared to our Victorian and Puritan ethics. It is also a significant change from the much more narrow permissiveness with affection standard. Finally, I should also stress that I was not inventing this HER sexual ethic. It was very much the ethic I had encountered in Sweden and increasingly found evidence for in America.

In this 1990 book, I was directly confronting the absolutism of those traditionalists who said we should reject teenage sex, homosexuality, and sex outside of marriage. I examined the evidence of trends all over the Western world toward HER sexual pluralism and argued that we are now in the midst of a new Western sexual revolution. Today's revolution concerns sexual attitudes and not so much sexual behavior. It is a more silent revolution. And as such it is only partially recognized but it is rapidly spreading in the Western world and we will increasingly become aware of it. This change will complete the sexual revolution that burst forth in the late 1960s.

I made a number of predictions in this 1990 book that I thought would occur before the end of the 1990s. In my 1997 revision of that book I checked them out. I had predicted lower teenage pregnancy rates, increased condom use, decreased HIV/AIDS rates, decreased rape rates, and greater tolerance for gays and lesbians (Reiss 1990, 234–236). By 1997, almost all these changes were clearly happening (Reiss 1997a, 13–14). I also had predicted that by the end of the 1990s our country would have explicitly endorsed the HER sexual pluralism ethic. The evidence from national surveys indicated that I was right on the money (Reiss 1997a). More recently, the American public's reaction to former president Bill Clinton's affair with Monica Lewinsky perhaps most clearly indicates that we as a nation have moved toward a more pluralistic approach to sexuality. Two-thirds of the public consistently said that no matter what they thought of Clinton's affair, they felt it was not the business of the public and he should remain in office. Few of us would have predicted such a supportive view for the privacy of extramarital sexual behavior that most Americans personally reject.

My 1990 book exemplified a much more direct legitimate role for sociological work in coping with social problems. My approach today surely does not deny the value of less prescriptive science and mainstream research work. Rather, I am validating the need to add a prescriptive dimension to our work and to develop a portrait of what we "could" be like, rather than focusing only on what we are like. This stance also incorporates elements of my views as a graduate student emphasizing the importance of what is significant and lowering the priority of focusing so heavily on what is precisely measurable. It is a reemphasis of my admiration for the work of people like Gunnar Myrdal (1944) who long ago was bold enough to suggest how we might resolve our "American dilemma" about race and equality. I sought to reclaim many of the powers that social scientists abdicated to the politicians. I am convinced that we in science need to work together with those in power to develop a common vision of how to change society and not give the "vision thing" completely over to the politicians. I fully believe that we can, with conscious effort, do this and still maintain fairness and balance in our scientific work.

Quo Vadis?

Surely sociology has played a major role in my personal life. It has helped me in my own ethical decisions. But I want to be very clear that I still firmly believe that our personal moral views should not be allowed to overwhelm our scientific integrity. I do not want to reproduce the biased stances taken by the 1953 family textbooks. The rigor, logic, empiricism, and openness to new ideas of science must be preserved and given top priority. But instead of seeking the impossible goal of being value free, we need to work harder toward the achievable goal of being value aware and value fair (Reiss 1993, 1997b, 1999). In that way, we become more likely to be aware of how our values can bias our work and thus more likely to consciously give priority to avoiding that pitfall. Instead of allowing our values and assumptions to unknowingly creep into our work or pretending that values have no impact on our work, we need to stress value awareness and value fairness so as to expose our values while working to prevent them from overwhelming our science.

I now feel that I have reached a much broader audience; not a mass audience, but a broad audience of educated Americans. Groups like Planned Parenthood responded enthusiastically to my 1990 book. I wrote the book without jargon and with no tables or graphs. I worked for eighteen months with my nephew Spence Porter, a New York playwright, who taught me how to add dramatic quality, use anecdotes, and make my ideas relevant to current events. My wife Harriet increased my awareness of the reader's needs and gave me deeper insights into a woman's perspective on gender equality issues. By the end, I had improved my ability to write in clear, plain, and interesting English, and I hope I have retained some of that in this autobiography.

Up until the mid-1980s, I believed that sexuality research and theory would do best by remaining as a subfield in the established disciplines of sociology, psychology, anthropology, biology, and such. But as I began to write for a more multidisciplinary audience, I changed my thinking. I realized that sexuality had been held back by being a subfield and often a low-ranked one, in other disciplines. I then began to argue for establishing a separate discipline of sexual science in order to advance the credibility and support for our work. Finally, in 1998 I persuaded both the Society for the Scientific Study of Sexuality and the American Association of Sex Educators, Counselors, and Therapists to form a joint task force to work to encourage the development of a Ph.D. program in sexual science at one of our major universities. Since I retired from the University of Minnesota in 1996, I had time to chair this task force. I wrote up a general outline of a Ph.D. program in sexual science that spelled out a multidiscipline and multiuniversity format for this new sexual science field (Reiss 1999, 258–267). In May 1999, our task force met at the Kinsey Institute at Indiana University and was successful in persuading the Kinsey Institute to adopt this perspective and work toward the establishment of a Ph.D. program in sexual science at Indiana University.

The first major step in this direction was taken by the Kinsey Institute when it succeeded in obtaining government and private grants that allowed it to bring in top sexual scientists from around the country to conduct its first summer institute in the summer of 2001. This is just the first step toward a separate Ph.D. in sexual science field, but it is a most important one. This planned Ph.D. program would be the first at a major American university and it will do much to establish the legitimacy and creditability of sexual science (Reiss 1999).

I am an optimist and I strongly believe that the field of sexual science will prosper in this new century. We can make the twenty-first century a time during which we as a people will increasingly learn how to reduce our problems with sexuality while at the same time learning how to achieve more of the rewards of sexuality. I believe that we are on the threshold of major advances in society, in sexual science, and in our personal lives. We can all play a role by supporting these important trends.

References

Burr, Wesley, Reuben Hill, Ivan Nye, and Ira L. Reiss, eds. 1979. *Contemporary Theories about the Family*. Vols. 1 and 2. New York: Free Press.

Coutu, Walter. 1949. *Emergent Human Nature*. New York: Knopf.

Deven, Fred, and Philip Meredith. 1997. "The Relevance of a Macrosociological Perspective on Sexuality for an Understanding of the Risks of HIV Infection." In *Sexual Interactions and HIV Risk: New Conceptual Perspectives in European Research*, ed. Luc Van Campenhoudt, Mitchell Cohen, Gustavo Guizzardi, and Dominique Hausser. London: Taylor and Francis.

Glass, Shirley P., and Thomas L. Wright. 1992. "Justifications for Extramarital Relationships: The Association Between Attitudes, Behaviors and Gender." *Journal of Sex Research* 29 (August): 361–387.

Hopkins, Kenneth Wu. 2000. "Testing Reiss's Autonomy Theory on Changes in Nonmarital Coital Attitudes and Behaviors of U.S. Teenagers: 1960–1990." *Scandinavian Journal of Sexology* 3 (December): 113–125.

Kephart, William M. 1961. Review of *Premarital Sexual Standards in America*, by Ira L. Reiss. *American Sociological Review* 26 (April): 294–295.

Murdock, George P., and Douglas R. White. 1969. "Standard Cross Cultural Sample." *Ethnology* 8 (October): 329–369.

Myrdal, Gunnar. 1944. *An American Dilemma*. New York: Harper.

Reiss, Ira L. 1956. "The Double Standard in Premarital Sexual Intercourse: A Neglected Concept." *Social Forces* 34 (March): 224–230.

———. 1957. "The Treatment of Pre-marital Coitus in 'Marriage and the Family' Texts." *Social Problems* 4 (April): 334–338.

———. 1960a. *Premarital Sexual Standards in America*. Glencoe, Ill.: Free Press.

———. 1960b. "Toward a Sociology of the Heterosexual Love Relationship." *Marriage and Family Living* 22 (May): 139–145.

———. 1964. "The Scaling of Premarital Sexual Permissiveness." *Journal of Marriage and the Family* 26 (May): 188–198.

———. 1965. "The Universality of the Family: A Conceptual Analysis." *Journal of Marriage and the Family* 27 (November): 443–453.

———. 1967. *The Social Context of Premarital Sexual Permissiveness*. New York: Holt, Rinehart and Winston.

———. 1971. *Family Systems in America*. New York: Holt, Rinehart and Winston. (2nd ed. 1976; 3rd ed. 1980.)

———. 1980. "Sexual Customs and Gender Roles in Sweden and America: An Analysis and Interpretation." In *Research in the Interweave of Social Roles: Women and Men*, ed. Helena Lopata. Greenwich, Conn.: JAI Press.

———. 1986. *Journey into Sexuality: An Exploratory Voyage*. New York: Prentice-Hall.

———. 1990. *An End to Shame: Shaping Our Next Sexual Revolution*. Amherst, N.Y.: Prometheus.

———. 1993. "The Future of Sex Research and the Meaning of Science." *Journal of Sex Research* 30 (February): 3–11.

———. 1997a. *Solving America's Sexual Crises*. Amherst, N.Y.: Prometheus.

———. 1997b. "An Introduction to the Many Meanings of Sexological Knowledge." In *International Encyclopedia of Sexuality*, ed. Robert T. Francoeur. Vol. 1. New York: Continuum.

———. 1998. "Reiss Male and Female Premarital Sexual Permissiveness Scales" and "Reiss Extramarital Sexual Permissiveness Scales." In *Handbook of Sexuality-Related Measures*, ed. Clive M. Davis, William L. Yarber, Robert Bauserman, George Schreer, and Sandra L. Davis. 2nd ed. Thousand Oaks, Calif.: Sage.

———. 1999. "Evaluating Sexual Science: Problems and Prospects." *Annual Review of Sex Research* 10: 236–271.

———. 2001. "Sexual Attitudes and Behavior." In *International Encyclopedia of the Social and Behavioral Sciences*, ed. Neil J. Smelser and Paul B. Baltes. Vol. 21. Oxford: Elsevier Science.

Reiss, Ira L., and Gary R. Lee. 1988. *Family Systems in America*. 4th ed. New York: Holt, Rinehart and Winston.

Reiss, Ira L., and Robert K. Leik. 1989. "Evaluating Strategies to Avoid AIDS: Number of Partners vs. Use of Condoms." *Journal of Sex Research* 26 (November): 411–433.

Reiss, Ira L., and Brent C. Miller. 1979. "Heterosexual Permissiveness: A Theoretical Analysis." In *Contemporary Theories about the Family*, ed. Wesley Burr, Reuben Hill, Ivan Nye, and Ira L. Reiss. Vol. 1. Glencoe, Ill.: Free Press.

Reiss, Ira L., Ron Anderson, and G. C. Sponaugle. 1980. "A Multivariate Model of the Determinants of Extramarital Sexual Permissiveness." *Journal of Marriage and the Family* 42 (May): 395–411.

Reiss, Ira L., Albert Banwart, and Harry Foreman. 1975. "Premarital Contraceptive Usage: A Study and Some Theoretical Explanations." *Journal of Marriage and the Family* 37 (August): 619–630.

Schwartz, Israel, and Ira L. Reiss. 1995. "The Scaling of Premarital Sexual Permissiveness Revisited: Test Results of Reiss's New Short Form version." *Journal of Sex and Marital Therapy* 21 (Summer): 76–86.

Suggs, David N., and Andrew W. Miracle, eds. 1993. *Culture and Human Sexuality: A Reader.* Pacific Grove, Calif.: Brooks/Cole.

Note

This appendix is a revision of "The Autobiography of a Sex Researcher: Short Version," by Ira L. Reiss, from *Marriage and Family Review* 31, nos. 1–2 (2001): 57–91. Copyright © 2001 by Haworth Press. Reprinted by permission of the publisher.